Band 21 · November 1985 **11**

isotopenpraxis

AKADEMIE-VERLAG

ISSN 0021-1915 Isotopenpraxis, Berlin 21 (1985) 11, 373–411

EVP 9,– M

An unsere Autoren

Die Zeitschrift veröffentlicht Beiträge zu folgenden Themen: Kontrolle und Analyse radioaktiver und stabiler Isotope, Anwendung radioaktiver und stabiler Isotope in allen Gebieten der Forschung und Technik, der Industrie, Landwirtschaft und Medizin, Herstellung radioaktiver und stabiler Isotopen sowie von markierten Verbindungen, kernelektronische Geräte für die Isotopenanwendung (Dickenmesser, Dichtemesser u. ä.) Detektoren für isotopentechnische Arbeiten und den Strahlenschutz in Isotopenlaboratorien, Meßtechnik und Strahlenschutz, Dosimetrie in Isotopenlaboratorien, kerntechnische Geräte für die Isotopenanwendung (Beton, Bestrahlungsanlagen u. ä.), Strahlenchemie, Radiochemie, ökonomische und perspektivische Fragen, Fragen der Weiterbildung.

1. Aufnahme von Manuskripten

Zur Veröffentlichung angenommen werden nur druckreife, noch nicht anderweitig publizierte Beiträge in deutscher, englischer oder russischer Sprache zum Themenbereich der Zeitschrift und zwar

Originalarbeiten
Kurze Mitteilungen
Aus der Praxis
Zusammenfassende Berichte
Tagungsberichte
Informationen.

Sind der Beitrag oder Teile daraus bereits veröffentlicht oder ist eine anderweitige Veröffentlichung vorgesehen, so muß dies der Redaktion mit genauer Angabe der Zweitveröffentlichung mitgeteilt werden. Mit der Annahme des Manuskriptes behält sich der Verlag das Recht der Vervielfältigung, Verbreitung und Übersetzung innerhalb der gesetzlichen Schutzfrist vor. Auszugsweiser Nachdruck oder photomechanische u. a. Wiedergabe sind an eine schriftliche Genehmigung des Verlages gebunden. Manuskripte, Zeichnungen und Photos von veröffentlichten Arbeiten werden nur auf den besonderen und bis zum Zeitpunkt der Veröffentlichung mitgeteilten Wunsch des Autors zurückgesandt.

Bei der unverlangten Einreichung von Manuskripten wird empfohlen, eine Befürwortung eines der Herausgeber oder Redaktionsausschußmitglieder beizufügen.

Manuskripte sind einzureichen bei der Redaktion, 1086 Berlin Leipziger Str. 3–4.

2. Form der Manuskripte

Die Autoren werden gebeten:

2 Manuskripte, eines davon in Original-Maschinenschrift, einseitig und mit 2zeiligem Abstand einzureichen; gut leserliche Photokopien sind zulässig; auf dem 1. Blatt (unter Weglassung der Bezeichnung)

den Titel der Arbeit
Vorname-Initialen und Namen sämtlicher Autoren
gegebenenfalls die Institution, in der sie tätig sind
eine kurze Zusammenfassung der Arbeit von höchstens 15 Zeilen
Keywords in alphabetischer Reihenfolge und
als Fußnote die Anschrift der Institution anzugeben;

wenn möglich, Zusammenfassung, Titel und Autoren **englisch** (bei deutschem oder russischem Manuskript) bzw. **deutsch** (bei englischem Manuskript) 2fach beizufügen;

Literaturhinweise (fortlaufend numeriert), Fußnoten, Tabellen, Abbildungen (in 2–3facher Veröffentlichungsgröße), Abbildungsunterschriften (getrennt von den Abbildungen mit Zeichenerklärungen, Quellenangaben, falls erforderlich, und gegebenenfalls mit einem Vermerk für besondere Wünsche bezüglich der Abbildungsgröße), private und Dienstanschrift der Verfasser mit akademischen Titeln und ausgeschriebenen Vornamen auf besonderen Blättern ans Ende des Manuskriptes zu stellen;

Wünsche für den Kolumnentitel (Seitenüberschrift) bekanntzugeben;

Absätze durch Einzüge von mindestens 3 Anschlägen kenntlich zu machen;

die Maßeinheiten des SI-Systems („Systeme International") zu verwenden;

Korrekturen im Manuskript unmittelbar an der zu korrigierenden Stelle anzubringen;

Formeln deutlich lesbar und möglichst groß einzutragen;

Farbabbildungen und nachträgliche Änderungen oder Ergänzungen in Abbildungen zu vermeiden;

hervorzuhebende Textstellen zu unterstreichen, und zwar halbfetten Satz mit einer geraden schwarzen Linie, kursiven Satz (Schrägschrift) mit einer blauen Wellenlinie (alle Eigennamen). Sperrungen gestrichelt;

besondere Buchstaben durch farbige Unterstreichung zu kennzeichnen, und zwar Formelzeichen in

Griechisch-Kursiv ($\alpha, \beta, \ldots$) rot,
Kursiv ($A, B, \ldots$) blau
halbfetter Kursiv ($\boldsymbol{A}, \boldsymbol{B}, \ldots$) 2 Linien blau (Vektoren),
halbfetter Grotesk ($\mathbf{A}, \mathbf{B}, \ldots$) 2 Linien violett (Tensoren);

bei Literaturangaben auf Vollständigkeit zu achten. In nachstehender Reihenfolge müssen angegeben sein:

a) bei *Zeitschriften*: Vorname-Initialen und Namen sämtlicher Autoren, Titel der Zeitschrift (nach „Periodica Chimica" oder möglichst wenig gekürzt), Bandnummer (unterstrichen), Jahreszahl (in Klammern), Heftnummer (nur bei Zeitschriften, die jedes Heft mit S.1 beginnen) sowie die Seitenzahl, auf der die zitierte Arbeit beginnt.

b) bei *Büchern*: Initialen der Vornamen und Name des Autors (Herausgeber werden durch den Zusatz „ed." gekennzeichnet), voll ausgeschriebener Titel des Buches in der Originalsprache, evtl. Bandnummer, voll ausgeschriebener Name des Verlages, Nummer der Auflage (außer der ersten), Erscheinungsort(e), Erscheinungsjahr (hier nicht in Klammern setzen!) und evtl. Kapitelnummer, Seitenzahl u. ä.

c) bei *Sammelbänden, Tagungsberichten*: Es soll zusätzlich der Titel des zitierten Beitrags angegeben werden. Die Seitenzahl muß unbedingt angegeben werden. Aus dem Zitat muß hervorgehen, ob es sich um einen Tagungssammelband oder um einen noch nicht veröffentlichten Vortrag handelt. Zwischen Tagungsort und Erscheinungsort des entsprechenden Tagungsberichts ist zu unterscheiden.

3. Korrekturen

Der Autor erhält zwei Umbruchabzüge, von denen einer möglichst schnell korrigiert an die Redaktion zurückgereicht werden soll. Auf Wunsch des Autors kann die Korrektur auch von der Redaktion gelesen werden.

Durch nachträgliche Manuskriptänderungen verursachte Korrekturkosten gehen, sofern sie 10% der Satzkosten übersteigen, zu Lasten des Autors.

Der Autor erhält 50 Sonderdrucke kostenlos. Weitere Sonderdrucke können bis zur Rückgabe der Korrekturabzüge auf Kosten des Autors bei der Redaktion bestellt werden.

Bezugsmöglichkeiten

Bestellungen sind zu richten

- **in der DDR** an den Postzeitungsvertrieb unter Angabe der Kundennummer des Bestellenden oder an den AKADEMIE-VERLAG BERLIN, DDR-1086 Berlin, Leipziger Straße 3–4, PF-Nr. 1233;
- **im sozialistischen Ausland** an eine Buchhandlung für fremdsprachige Literatur oder an den zuständigen Postzeitungsvertrieb;
- **in der BRD und Berlin(West)** an eine Buchhandlung oder an die Auslieferungsstelle KUNST UND WISSEN, Erich Bieber OHG, Wilhelmstraße 4–6, D-7000 Stuttgart 1;
- **in den übrigen westeuropäischen Ländern** an eine Buchhandlung oder an die Auslieferungsstelle KUNST UND WISSEN, Erich Bieber GmbH. Dufourstraße 51, CH-8008 Zürich;
- **im übrigen Ausland** an den Internationalen Buch- und Zeitschriftenhandel; den Buchexport Volkseigener Außenhandelsbetrieb der Deutschen Demokratischen Republik, DDR-7010 Leipzig, Postfach 160; oder an den AKADEMIE-VERLAG BERLIN, DDR-1086 Berlin, Leipziger Straße 3–4.

Zeitschrift „Isotopenpraxis"

Herausgeber: Zentralinstitut für Isotopen- und Strahlenforschung der Akademie der Wissenschaften der DDR.
Verlag: Akademie-Verlag Berlin, DDR-1086 Berlin, Leipziger Straße 3–4, PF-Nr. 1233; Fernruf: 2236221 oder 2236229; Telex-Nr.: 114420;
Bank: Staatsbank der DDR, Berlin, Kto.-Nr.: 6836-26-20712.
Chefredakteur: Prof. Dr. H. Hübner, Leipzig.
Anschrift der Redaktion der Zeitschrift Isotopenpraxis: Prof. Dr. H. Hübner, DDR-7010 Leipzig, Karl-Marx-Platz, Hauptpostamt, PSF 370; Fernruf: 281018.
Veröffentlicht unter der Lizenznummer 1527 des Presseamtes beim Vorsitzenden des Ministerrates der Deutschen Demokratischen Republik.
Gesamtherstellung: VEB Druckerei „Thomas Müntzer", DDR-5820 Bad Langensalza.
Erscheinungsweise: Die Zeitschrift „Isotopenpraxis" erscheint monatlich. Die 12 Hefte eines Jahrganges bilden einen Band. Bezugspreis je Band 240,– DM zuzüglich Versandspesen; Preis je Heft 20,– DM. Der gültige Jahresbezugspreis für die DDR ist der Postzeitungsliste zu entnehmen.
Bestellnummer dieses Heftes: 1079/21/11.

AN (EDV) 12831
00900

Isotopenpraxis 21 (1985) 11, S. 373 bis 374

Georg von Hevesy — sein 100. Geburtstag

W. Göbel

(Institut für Technologie der Polymere der AdW der DDR)[1])

Keywords

hafnium; history; phosphorus 32; radioisotopes; tracer techniques

Georg von Hevesy wurde unter dem Namen *György József Bisicz* am 1. 8. 1885 in Budapest als Sohn des damaligen Direktors der Oberungarischen Berg- und Hüttenwerks-AG Budapest, *Ludwig David Bisicz*, geboren. 1895 wurde letzterer in den erblichen Adelsstand erhoben. Der Adelsname lautete *Bisicz de Heves*. Ab 1906 wurde die Erlaubnis zur Führung des Namens *Hevesi de Heves* gegeben, woraus in der Verdeutschung schließlich der uns wohlbekannte und anläßlich seines diesjährigen 100. Geburtstages zu ehrende *Georg von Hevesy* wurde.

Georg von Hevesy

Hevesy studierte Physik, Mathematik und Chemie in Budapest, Berlin und Freiburg im Breisgau, wo er 1908 bei *G. Meyer* über Wechselwirkungen zwischen Natrium und geschmolzenem Natriumhydroxid promovierte. Danach war er zwei Jahre lang als Assistent bei *R. Lorenz* an der ETH Zürich tätig. Hier besuchte er auch die Experimentalvorlesungen des jungen *A. Einstein*. Es folgte ein kurzer Arbeitsaufenthalt bei *F. Haber* in Karlsruhe. *Haber* übertrug *Hevesy* die Aufgabe festzustellen, ob die Oxidation von geschmolzenem Zink von Elektronenemission begleitet wird. Dazu war es aber unerläßlich, Strahlungsmeßmethoden zu beherrschen, über die an *Habers* Institut keinerlei Erfahrungen vorlagen. Deshalb wurde *Hevesy* 1911 zu *E. Rutherford* nach Manchester entsandt, um sich zunächst dieser Problematik anzunehmen.

Rutherford hatte schon in den ersten Jahren unseres Jahrhunderts eine beträchtliche Menge radioaktives Blei-(IV)-chlorid als Abfall der Radiumherstellung erhalten. Er untersuchte damals die Strahlung, die dieses Material emittierte und deren Urheber das damals noch nicht identifizierte RaD war. Er forderte *Hevesy* auf, das RaD von der großen Menge des inaktiven Bleis abzutrennen. Nach langer und mühseliger Arbeit erkannte dieser, daß die Abtrennung durch chemische Methoden nicht möglich ist und daß sich RaD und Pb wie „praktisch chemisch identische" Stoffe verhalten. (Der Begriff Isotopie wurde erst 1913 von *F. Soddy* (Nobelpreis 1921) eingeführt.)

Der Mißerfolg führte *Hevesy* zu folgender Überlegung: wenn man RaD und Pb nicht voneinander trennen kann, dann ist es möglich, durch RaD-Zugabe zu inaktivem Blei bzw. zu dessen Verbindungen dieses Element radioaktiv zu „markieren".

Im November 1912 kam *Hevesy* nach Wien. Noch 1913 entwickelte er hier gemeinsam mit *F. Paneth* im Wiener Radium-Institut der Österreichischen Akademie der Wissenschaften die Leitisotopen- oder Tracer-Methode. Im gleichen Jahr habilitierte er sich an der Universität Budapest.

1912 erkannte *Hevesy* auch die Abhängigkeit der Gültigkeit der *Nernst*schen Gleichung vom Grad der Bedeckung der Elektrodenoberfläche mit den abgeschiedenen Atomen der Nuklidspuren. *Hevesy* und *Paneth* zeigten 1914 auch die Möglichkeit der Untersuchung elektrochemischer Prozesse einer radioaktiven Spur. Mit der Zeit wurden Methoden entwickelt, um auch in diesen Fällen den Wert des Elektrodenpotentials zu ermitteln.

Im 1. Weltkrieg leitete *Hevesy* als k. u. k. Ingenieur-Offizier elektrochemische Raffinerien in Ungarn. Während dieser Zeit arbeitete er auch wissenschaftlich auf seinem eigentlichen Arbeitsgebiet weiter.

Nach Kriegsende hielt er bis 1920 Vorlesungen an der Universität Budapest. Das ihm 1920 angebotene Ordinariat für physikalische Chemie an der Universität Budapest nahm er nicht an, sondern folgte einer Einladung von *N. Bohr* und ging nach Kopenhagen. Im gleichen Jahr gelang *Hevesy* mittels radioaktiver Markierung die experimentelle Untersuchung der Selbstdiffusion und der Austauschreaktionen zwischen Atomen desselben Elements in verschiedenen Verbindungen. Auch die Untersuchungen an zahlreichen biologischen Objekten, die bald einen immer größer werdenden Umfang annehmen sollten, wurden erst durch die Leitisotopenmethode möglich.

In Kopenhagen arbeitete *Hevesy* zunächst gemeinsam mit *J. N. Brønsted* an der teilweisen Trennung der Isotope des Chlors und des Quecksilbers. Hier entdeckte er auch 1922 gemeinsam mit *D. Coster* das Element Hafnium, benannt nach dem lateinischen Namen von Kopenhagen,

[1]) Anschrift: DDR-8010 Dresden, Hohe Str. 6

hafnia. Als *Bohr* am 10. Dezember 1922 seinen Nobelvortrag hielt, teilte er die Entdeckung des Hafniums mit. Es war aus einem norwegischen Zirkonmineral angereichert und mittels Röntgenspektralanalyse nachgewiesen worden. Berechnete und experimentell gefundene Wellenlängen befanden sich in guter Übereinstimmung.

Es folgten u. a. Arbeiten zur Trennung der Isotope des Kaliums und erste Untersuchungen mit radioaktiven Isotopen an lebenden Organismen.

1926 folgte *Hevesy* dem Ruf auf das Ordinariat für physikalische Chemie der Universität Freiburg/Breisgau. Hier arbeitete er auf Teilgebieten der Geo- und Kosmochemie und entwickelte die Röntgenfluoreszenzanalyse. Auch gelang ihm in dieser Zeit der Nachweis, daß das radioaktive Isotop des Kaliums jenes mit der Atommasse 40 ist.

Die sehr empfindliche Methode der Benutzung von Rückstoßatomen als radioaktive Indikatoren wurde von *Hevesy* und *W. Seith* 1929 aufgefunden. 1931 entwickelte *Hevesy* gemeinsam mit *R. Hobbie* die Isotopenverdünnungsmethode. Ihr Prinzip ist von ihnen erstmalig zur mikroanalytischen Bestimmung von Bleispuren angewandt worden.

1932 versuchten sich *Hevesy* und *M. Pahl* am natürlichen Samarium, scheiterten aber noch beim Versuch der richtigen Zuordnung der beobachteten α-Aktivität.

Nach dem Machtantritt der Faschisten in Deutschland verließ *Hevesy* 1934 Freiburg und ging zurück nach Kopenhagen, wo er von 1935—1943 als Ordinarius am Institut für theoretische Physik der Universität wirkte. Hier widmete er sich dem Studium biologischer und medizinischer Probleme, die er mit Hilfe der Tracer-Methode zu lösen versuchte. Die Reaktion $^{32}S(n,p)^{32}P$ wurde 1935 von *O. Chiewitz* und *Hevesy* ursprünglich mit dem Ziel der Herstellung von Radiophosphor mit hoher spezifischer Aktivität untersucht. Mit dem so gewonnenen ^{32}P gelang dann der Nachweis, daß die mineralischen Bestandteile des Skeletts dauernd erneuert werden. Es konnten später weitere dynamische Prozesse im Organismus nachgewiesen werden.

1936 entwickelten *Hevesy* und *Hilde Levi* die Aktivierungsanalyse. Eine Möglichkeit ihrer Nutzung demonstrierten sie bei der Bestimmung von Gadolinium in einem Lanthaniden-Präparat.

Hevesy flüchtete 1943 während der faschistischen Okkupation Dänemarks nach Schweden. Von 1943—1956 wirkte er am Stockholmer Universitätsinstitut für organische Chemie als wissenschaftlicher Mitarbeiter.

1943 wurde ihm der Nobelpreis, insbesondere für seine Arbeiten über die Anwendung der Isotope als Indikatorsubstanzen bei der Erforschung chemischer Prozesse, oder, wie es im authentischen englischen Text heißt, „... for his work on the use of istotopes as tracers in the study of chemical processes ... “, zuerkannt.

Auch noch nach 1956 bis kurz vor seinem Tode, der ihn am 5. 7. 1966 in einer Freiburger Klinik, in der er von einer schweren Krankheit zu genesen hoffte, ereilte, war er wissenschaftlich tätig. Seine letzten Arbeiten galten wiederum biologischen, biochemischen und medizinischen Problemen, bei denen er die Tracer-Methode einsetzte.

Hevesy war Auswärtiges Mitglied der Royal Society, zwölffacher Ehrendoktor und Träger vieler anderer internationaler Auszeichnungen. Die Zahl seiner wissenschaftlichen Veröffentlichungen beträgt nahezu 400 [1—6].

Aus Anlaß des 100. Geburtstages von *Georg von Hevesy* fand am 10. 4. 1985 ein Gedenkkolloquium in Dresden statt (Tagungsbericht vgl. dieses Heft, S. 404), das vom Vorsitzenden des Fachverbandes Chemische Analytik der Chemischen Gesellschaft der DDR, Prof. Dr. *Gerhard Ackermann*, Freiberg, eröffnet wurde. Er konnte auch ungarische Kollegen begrüßen. Das Kolloquium war ein beredter Beweis dafür, daß *Hevesys* Werk unvergessen ist und daß, auf seinen Erkenntnissen aufbauend, ständig neue Methoden und Anwendungen entwickelt bzw. gefunden werden können.

W. Göbel

Eingegangen am 23. 4. 1985

Literatur

[1] *J. C. Poggendorff*, Biographisch-Literarisches Handwörterbuch der exakten Naturwissenschaften, Band VII b, Teil 4, 1. Lieferung, Berlin: Akademie-Verlag 1971, S. 1972
[2] Neue Deutsche Biographie, Band 9. Berlin: Duncker & Humbolt Verlag 1972, S. 61
[3] *F. Szabadváry*, J. radioanalytical Chem. (Budapest) **1** (1968) 97
[4] *H. Levi*, Nuclear Physics A **98** (1967) 1
[5] Nature (London) **211** (1966) 464
[6] *V. Majer*, Grundlagen der Kernchemie, Leipzig 1982, S. 126, 324 531, 532, 816, 827, 841, 901, 919, 929

Isotopenpraxis 21 (1985) 11, S. 375 bis 379

Originalarbeiten

Application of Ionizing Radiation for Production of Drugs, Vaccines, and Biochemicals*)

W. Nordheim, S. Bräuniger, G. Petzold

(Institute for Applied Virology, Berlin, GDR)[1])

J. Reinhardt, M. Bär

(Academy of Sciences of the GDR, Central Institute of Isotope and Radiation Research, Leipzig, GDR)[2])

K.-Ch. Bergmann

(Research Institute on Lung Diseases and Tuberculosis, Berlin, GDR)[3])

^{60}Co-gamma-rays can be used efficiently for sterilizing products, such highly refined ones as drugs, vaccines, biochemicals. Examples of advanced stage application are: sterilization of enzymes used for food production and of supplements for drug manufacture. As economical aspects are of growing importance also for production of vaccines and antigens for diagnostic purposes, methods should take into account quantitative factors and qualitative biological properties. Therefore cautious inactivation of microorganisms incl. viruses is demonstrated. Special problems of irradiation are discussed.

^{60}Co-Gamma-Strahlung kann wirkungsvoll eingesetzt werden, um Produkte mit hohem Veredlungsgrad wie Medikamente, Impfstoffe und Biochemikalien zu sterilisieren. Beispiele einer Anwendung in fortgeschrittenem Stadium bieten die Sterilisation von Enzymen für die Lebensmittelproduktion und von Hilfsstoffen für die pharmazeutische Industrie. Da ökonomische Aspekte von zunehmender Bedeutung auch für die Impfstoffherstellung und die Produktion diagnostischer Antigene sind, sollten die Verfahren quantitative Faktoren und biologische Eigenschaften berücksichtigen. In diesem Zusammenhang wird die schonende Inaktivierung von Mikroorganismen einschließlich Viren demonstriert. Spezielle Probleme der Bestrahlung stehen zur Diskussion.

Keywords

antibiotics; bacillus; drugs; enzymes; gamma radiation; irradiation; sterilization; vaccines; viruses

Introduction

Production of vaccines, drugs, pharmaceuticals and biochemicals — including enzymes — is possible only with great endeavours and a high material input, so that the manufactured products are rather expensive. Therefore it is a disadvantage, if

- the quality of these products is reduced during manufacturing or as final products by undesired microbial or viral contaminants
- there are great losses in yield (in the process of production).

This implies in the first case that the product will be released only with special permissions or the release is altogether interdicted. In the second case a high input of material is demanded.

There are two approaches to overcome this drawback:

- The first consists in controlling microbial and viral contaminations or infections,
- the second consists in mild inactivation of viruses and microorganisms in order to manufacture high-quality products.

*) Plenary lecture presented at the "Third Working Meeting on Radiation Interaction", $24^{th}-28^{th}$ September 1984, Leipzig, GDR

Addresses: [1]) 1190 Berlin, Britzer Str. 1–3; [2]) 7050 Leipzig, Permoserstr. 15; [3]) 1115 Berlin, Karower Str. 11

The first approach

If already put into force, requirements for absence of microbial contamination differ, varying from the sterile state to a content of several thousands of nonpathogenic microorganisms. So far, however, aseptic methods as well as conventional degermination or sterilization have not succeeded in solving the problem. Chemical and thermal sterilization procedures often bring about a partial destruction or changes in the active principle of substances.

The rather difficult aseptic method can be obviated to a certain extent, if sterilization or reduction of the number of microorganisms is achieved by other methods at the end of the process. But this does not imply that the rules of good manufacturing practice are allowed to be infringed upon. That is to say that on the technical side enormous efforts have to be made to achieve a low number of germs or to obtain a germ-free product.

It is more difficult, however, if the raw material is heavily contaminated or if a strain of microorganisms is used for producing a drug or a biochemical product.

The second approach

For the production of inactivated vaccines or special diagnostic preparations it is necessary to inactivate the viruses resp. the microorganisms that have been grown for this special purpose. Special attention has to be paid to the process of inactivation in order not to destroy the antigenic and immunogenic properties which constitute the value of the product. At present conventional chemical and physical methods fraught with many disadvantages are used in the practice. Chemical substances used for inactivation might remain in the final product and contribute to an additional undesired reactogenicity.

The postulate of the two approaches is in any case inactivation of viruses and microorganisms.

One way to circumvent the disadvantages would be the use of irradiation, esp. gamma radiation. Although ionizing radiation can not be regarded as a remedy for all problems, our efforts have shown that the results are promising.

Results and discussion

We used gamma irradiation to achieve two aims:

1. Mild inactivation of viruses and microorganisms for obtaining products of value, e. g. vaccines
2. Degermination or at least reduction of viral and microbial contamination without destruction of the components containing the active principle

In all these cases it is of great advantage, that the antigenic and immunogenic components as well as special proteins, esp. enzymes, resp. biopolymers are more resistant against irradiation than nucleic acids responsible for replication.

Inactivation of viruses, bacteria, helminthes and parasitic protozoa

Effects of radiation can be direct or indirect. The virucidal or microbicidal effect of ionizing radiation is enhanced markedly by the reaction of water radiolysis products. This indirect effect of radiation — i.e. the reaction of products of water radiolysis with important biologically active substances — can be used methodically in many processes — not only for mild inactivation of viruses resp. microorganisms, but also for making the radiation input costs more effective.

Viruses exhibit the greatest resistance against ionizing radiation. Here we present data from the literature (Tab. 1) [1]. The following table contains data about bacteria. Yeasts and moulds belong into this range.

Tab. 1. Parameters for inactivation (examples)

Viruses max. 120 kGy *)	Bacteria max. 35 kGy *)
Vaccinia virus	Salmonella typh.
Poliomyelitis virus	Shigella flexneri etc.
FMD virus	Brucella abortus etc.
Influenza virus	Anaplasma marginale
Mumps virus	Erysipelothrix rhusiopathiae
Rabies virus	Listeria monocytogenes
Leukemia virus	Vibrio cholerae
Rubella virus	Leptospira icterohaemorrhagiae

*) radiation dose (literature)

Last of all are the helminthes and parasitic protozoa (Tab. 2) [1].

The following data have been up-dated by using the results of our experiments (Tab. 3).

Thus viruses are inactivated with a radiation dose of about 15 kGy.

Under certain conditions microorganisms can be inactivated already with ≦10 kGy as was shown e.g. in our experiments with Bacillus subtilis, Bacillus pumilus or Endomycopsis bispora.

Tab. 2. Parameters for inactivation (examples)

Helminthes max. 1,6 kGy *)	Parasitic Protozoa max. 1,3 kGy *)
Schistosoma bovis	Trypanosoma cruzi
Schistosoma mansoni	Plasmodium berghei
Ancylostoma caninum	Plasmodium gallinaceum
Oesophagostomum columbianum	Babesia divergens
Cooperia oncophora	Babesia bigemina etc.
Ostertagia ostertagi	Theileria parva
Dictyocaulus viviparus	Eimeria necatrix
Dictyocaulus filaria	Toxoplasma gondii

*) radiation dose (literature)

Tab. 3. Radioresistance: 5 virus species

Virus species	Dose for complete inactivation
Pseudorabies virus	10,5 kGy
Fowl pox virus	14,0 kGy
Newcastle disease virus	20,5 kGy
Reovirus 1	19,0 kGy
Influenza virus	15,0 kGy
Lyophilisates: increased radiation dose	

Drugs, vaccines and biochemicals

About the use of ionizing radiation for preparing drugs, vaccines and biochemicals an extensive study devised to be an orientation in this field, has been made in cooperation between the Institute for Applied Virology and the Central Institute of Isotope and Radiation Research. The study is published in the "ZfI-Mitteilungen" [1].

More than 400 drugs and other substances have been summarized, from adrenalin to citric acid.

Irradiation of drugs, biochemicals and auxiliary reagents is more advantageous in a solid state. In 65—70% of the probes no or only minimum decomposition was noted.

It has to be mentioned, however, that the doses of radiation — some being 1 MGy — were much too high. In the aqueous phase decomposition is enhanced with an equivalent irradiation dose and it increases with growing irradiation doses. Therefore irradiation should be done separately for solid-phase and aqueous-phase preparations. In these cases the use of radioprotective substances should be considered more often. Enzymes and other preparations containing protein often have a different reaction.

Taking into account the data obtained from the literature as well as from own tests about the necessary doses for inactivating microorganisms, the conclusion can be drawn that sterilization with radiation is a technical possibility for a large number of drugs, biochemicals and auxiliary reagents.

The level of degermination of drugs and auxiliary substances has been defined within a rather wide range depending on the kind of use. Only for injectible preparations the decision is unambiguous: these substances have to be definitely sterile.

As for vaccines, the problem is not simply to kill any foreign germs but to inactivate the manufactured microorganisms resp. viruses without damaging the antigens. As the direct and indirect radiation effects generally concern nucleic acids, the use of radiation technology for producing vaccines is especially advantageous.

The data published in the literature confirm that the results obtained in vaccinating with preparations or microorganisms inactivated by irradiation are comparable with conventional inactivated vaccines [1].

The results proved to be insufficient only if inactivation had been achieved by a "hard" irradiation, i.e. too high doses, without using temperature control and/or stabilizators.

From all this the conclusion can be drawn that ionizing radiation is a modern means for inactivation used in the production of inactivated vaccines and other inactivated preparations, but also for certain steps in the process of production of live vaccines. It could help to implement certain steps in the production process otherwise not possible due to economical or technical problems (e.g. total degermination, longer stability or selective inactivation of mycoplasmas).

Our position concerning investigation and development in 1984

It has to be stressed that we have achieved inactivation with essentially smaller doses of radiation than mentioned generally in the literature [1]. In the process of irradiation the dose is the primary economic parameter, and that is why reducing the doses is a means to make irradiation processes more economic.

Drugs

A special feature of our studies consisted in experiments for sterilization of "fatty oils" (caster oil, sunflower oil) used as drug carriers for manufacturing eye ointments which contained atropin, pilocarpin, scopolamin or penicillin. Bacillus pumilus was chosen as a test microorganism, for it is very resistant to radiation.

It was of great interest that already 4 kGy were sufficient for total disinfection, but without essential changes in alkali colour number, hydroxyl and peroxid number, saponification value and optical rotation.

Just now we are investigating the sterilization of gelatin, hydroxyethyl cellulose, sulfacetamid-sodium, some hormone preparations, certain steroids, antibiotics, pollen and vegetative mucoid extracts, e.g. camomile, as well as standard feed for "SPF-animals" used for drug control.

Mycoplasmas

Another problem is the killing of mycoplasmas in sera, esp. fetal calf sera, and in other special substrates, without interfering with the quality of these sera that are to be used e.g. to grow cells [2]. Sera for growing cells tolerate up to 25 kGy without damage. Simultaneously microorganisms can be inactivated with this dose. — An appropriate dose for inactivation of Acholeplasma laidlawii is 3 kGy [2].

Vaccines and diagnostic preparations

Now we present the results obtained in inactivating viruses to be used as vaccines and as diagnostic preparations.

The inactivation in the production process is for the most part carried out with chemical (e.g. formalin) or thermal treatment. This is valid for human and veterinary vaccines.

The main drawbacks are changes in antigenic protein, immunogenicity and safety, to mention only a few.

Following strains of influenza viruses have been subjected to irradiation inactivation procedures:

Type A — A/PR/8/34
— X-73 (A/PR/8/34 × A/Bangkok/1/79)
Type B — B/Singapore/222/79

Radiation doses were in the range between 10 and 20 kGy. No clear statistical evidence was obtained in favour of a decrease in activity of the two surface components of influenza virus, i.e. hemagglutinin and neuraminidase, among the tested strains after treatment with increasing radiation doses (Tab. 4).

Tab. 4. Activity decrease of influenza virus antigens after irradiation (mean values in p.c.)

Dose, kGy	Hemagglutinin	Neuraminidase
10	6	2,4
12,5	11	10,1
15	11	10,3
20	10	7,0

Tab. 5. Parameters for inactivation influenza viruses

	A/PR/8/34 and X-73		B/Singapore/222/79
Dose, kGy	12		12,5
Dose/h, kGy	8		8
Temp., °C	15 – 25		15 – 20
HA, µg/ml	–	200 – 600	–

Tab. 6. D_{10}-values for influenza viruses (virus-containing allantoic fluid)

Strain	Mean value, Gray
A/PR/8/34	1740 ± 150
X-73	1580 ± 270
B/Singapore/222/79	1770 ± 140

The range where radiation inactivation proved to be least damaging for protein has been determined for selected influenza viruses (Tab. 5). The $dose_{10}$ values were calculated (Tab. 6).

Testing for a possible reactivation has been extended to one year, without detecting any reactivation after treatment with only 12—16 kGy.

For comparison of influenza vaccines inactivated with radiation and with formalin or thermal treatment we present two tests. At first a monovalent vaccine (Tab. 7).

Tab. 7. Influenza virus vaccine — monovalent H3N2 (A/Dresden/1/79)

	Complete virus (conventionally inactivated)	Complete virus (radiation-inactivated)
HA	20,5	25,5
NA	12,3	17,3
Protein	312	290
Al	1,93	1,93
Antibody titer (HIT; mouse)	I. 1:128 II. 1:256	I. 1:128/256 II. 1:256/512

HA = Hemagglutinin, µg/ml
NA = Neuraminidase activity, µg NANA/50 µl
Protein, µg/ml
Aluminium, mg/ml

After chemical treatment with formalin the activity of the neuraminidase is still relatively high, though often the activity is reduced by 50% or more.

The next table presents the data obtained with a bivalent test vaccine (Tab. 8). Also in this case the irradiated vaccine is positive.

Tab. 8. Influenza virus vaccine – bivalent H3N2/H1N1 (NIB-4/NIB-6)

	Formalin-inactivated		Radiation-inactivated	
HA	H3: 36	H1: 13	H3: 47	H1: 16
NA	0,5		7,1	
Protein	200		194	
Al	0,95		0,39	
Antibody titer (HIT; guinea-pig)	NIB-4	NIB-6	NIB-4	NIB-6
	1:29,5	1:22	1:40	1:27

HA = Hemagglutinin, µg/ml
NA = Neuraminidase activity, µg NANA/50 µl
Protein, µg/ml
Aluminium, mg/ml

Prospects are exceedingly good for oral administration, as the requirements are less stringent, because these vaccines "can be regarded as food components".

Oral influenza vaccine would be an example. – For this purpose we prepared inactivated influenza virus vaccines.

The preparation is a simple procedure:

Virus reproduction
|
Virus purification resp. concentration
|
Inactivation (^{60}Co)
|
Manufacture
|
Vaccine for oral application

Hemagglutinin content and neuraminidase activity: Unchanged after irradiation (12 kGy)

Several modifications of influenza vaccines incativated with radiation have been tested in animals [3].

As compared to the parenteral group oral immunization with inactivated vaccines leads to higher secretory antibody response (Tab. 9). The antibody response correlates with protection against lethal virus challenge by aerosol characterized by a low cell yield and virus titer in the lung and high survival rate. It can be concluded that inactivation by irradiation can be used for production of oral influenza vaccines.

Tab. 9. Comparison of subcutaneous and oral immunization[1]) with inactivated influenza vaccine[2])

	Reciprocal mean antibodies titers			
	Lung lavage fluid		Serum	
Group	EIA[3])	LN[3])	EIA[3])	LN[3])
Control	<1	<4	<10	<20
Subcutaneous	1	<4	320	640
Oral	4	8	80	640

[1]) All groups (NMRI-mice) 2 × 25 µg HA within 7 days
[2]) A/PR/8/34 inactivated with 10 kGy
[3]) IgG-specific enzyme-immuno-assay and lectin-neuraminidase-inhibition test

Interesting results have been obtained also with the irradiation of test antigens [4]. Following test antigens have been investigated:

- Influenza virus type A
  strain NIB-6
- Influenza virus type B
  strain B/Singapore/222/79
- Mumps virus
  strain Enders

Noteworthy is the fact that – in comparison with the vaccine production – rather small doses, e.g. 500 Gy, proved to be sufficient. Antigen content and specificity were unchanged.

Enzymes

At last we present data concerning the microbial synthesis of enzymes as a practical example.

We obtained good results in sterilizing microbial enzyme solutions by gamma irradiation.

Following requirements have been established for using radiation:

- Elimination of disadvantages of conventional procedure (= high yield and quality)
- Enzymes – also for cold application
- Reduction of technical installation costs (= discontinuous operation)
- Economical use of ^{60}Co; depth effect
- No special temperature control

It has been our aim to use a radiation dose as low as possible. On account of many experiments following parameters have been adopted:

- Radiation dose 6000–9000 Gray
- Temperature 5–40 °C
- Preferentially aqueous preparations (>25% H_2O)
- Germ content: Reduction e.g. 10^7/ml → <10^2/ml

These parameters ensure that the germ counts required in the standards are not exceeded.

The plant in question manufactures the following enzyme preparations:

1. A complex preparation that is manufactured by submerse aerobic fermentation by Bacillus subtilis as the production strain. Obtained are β-glucanase as main activity and neutral protease and alpha-amylase as additional activities.
2. In addition a glucoamylase preparation is produced – also in submerse aerobic fermentation by Endomycopsis bispora.
3. An alpha-amylase preparation with high activity is planned to be manufactured with Bacillus subtilis in a modified aerobic fermentation.

After fermentation the culture suspension containing the enzymes are processed (or purified) and irradiated.

In addition these experiments showed that extracellular enzymes in a liquid phase tolerated even radiation doses of up to 20–25 kGy without losses in enzyme activities.

The irradiated enzymes being valuable biocatalysts are used for manufacturing food and – to a certain extent – drugs.

Classified as foreign substances these enzymes are subject to the regulations of the food act and the drug act and have to fulfil certain requirements concerning innocuity – independent of the microquantities used in the final product.

Tests for mutagenicity carried out so far did not provide evidence for changes due to irradiation.

Testing of irradiated and non-irradiated enzyme preparations with infrared spectroscopy proved to be interesting. No differences could be recognized in the bands, esp. those of carbonyl.

These results show that irradiation is a method for degermination of enzyme preparations that could be used for different purposes, e.g. for manufacturing food.

Irradiation has certain advantages, if compared with other methods for reducing the number of microorganisms or for sterilizing (e.g. with membrane filtration, thermal treatment or with chemicals) – these consist esp. in a higher yield and better safety in processing.

General conclusion

Certainly new questions will emerge in the course of implementation of these procedures, e.g. what kind of changes will occur in the products, wether so-called radiotoxins are formed under the conditions chosen by us and what potential influence upon the human organism might be seen following application of these products — either immediately as vaccines and drugs or as products manufactured with enzymes.

The methods chosen are legally admissible. The doses we use are in accordance with correspondent regulations [5].

Therefore it is not the proving of safety that is so difficult, but objections against use of irradiation, based on the association of irradiation with radioactivity, resulting in emotional objections against any kind of nuclear technique.

Introduction of the Cobalt-60 gamma irradiation technique in the biochemical and microbiological industries is not only kind of "spade-work", but will also bring about high economical results [6—13].

Received April 3, 1985

References

[1] *W. Nordheim, S. Bräuniger, B. Kirsch, H. Kotowski, D. Teupel*, ZfI-Mitteilungen 1984

[2] *E. Bender, J. Fritsche, W. Nordheim, M. Bär*, Third Working, Meeting on Radiation Interaction, Leipzig 1984

[3] *K.-Ch. Bergmann, K. Noack, H. Tischner, W.-D. Pohl, W. Nordheim, S. Bräuniger, G. Petzold*, Third Working Meeting on Radiation Interaction, Leipzig 1984

[4] *W. Nordheim, S. Bräuniger, G. Starke, G. Petzold, B. Kirsch, R. Nordheim*, (in Zusammenarbeit mit *M. Ladach*), 2. Tagung der Sektion Virologie der Gesellschaft für Mikrobiologie und Epidemiologie der DDR, Gera 1983, S. 64

[5] Gesetzblatt der DDR I Nr. 11 S. 151, 1984; AB-DDR 1983

[6] *W. Nordheim, H. Weise, M. Bär, G. Starke, H. Voigt, R. Nordheim, M. Wölbing*, DD WP 142058

[7] *W. Nordheim, M. Bär, G. Starke, S. Bräuniger, G. Petzold*, DD WP 146147

[8] *W. Nordheim, J. Neubert-Hils, G. Starke, S. Bräuniger, A. May, G. Petzold, Chr. Liebe*, DD WP 148721

[9] *W. Nordheim, H. Weise, M. Bär, G. Starke, H. Voigt, R. Nordheim, M. Wölbing*, DD WP 153892

[10] *W. Nordheim, H. Weise, H. Voigt, M. Wölbing, M. Bär, S. Bräuniger, R. Nordheim*, DD WP C 12N/226 915/8

[11] *K.-Ch. Bergmann, W. Nordheim, S. Bräuniger, K. Noack, H. Tischner, W.-D. Pohl, H. Schäfer, M. Bär, G. Starke, G. Petzold, B. Kirsch, D. Teupel, R. Nordheim*, DD WP 213351

[12] *K.-Ch. Bergmann, K. Noack, W. Nordheim, M. Bär, R. Nordheim*, DD WP C 12N/248 855/3

[13] *K.-Ch. Bergmann, W. Nordheim, S. Bräuniger, K. Noack, J. Reinhardt, G. Petzold, B. Kirsch, D. Teupel, R. Nordheim*, DD WP A 61 k/269 629 6

Isotopenpraxis **21** (1985) 11, S. 379 bis 383

Application of Low-Energy Electron-Beam Curing in Plastics Processing and Coating Technologies*

T. Czvikovszky

(Research Institute for the Plastics Industry)[1]

Industrial radiation processing using low-energy electron beam (EB) offers high-speed technologies for converting (chemically) thin layers of plastics, coatings, varnishes, printing inks etc., on different surfaces. Overviewing the main features of the corresponding technique, machines and materials used here, the actual state of chemistry and technology of EB-curable polymeric systems will be discussed on some examples taken from the plastics processing, paper-converting, wood-panel coating and other industries.

Die Verwendung nieder-energetischer Elektronenstrahlen bei industriellen strahlenchemischen Prozessen bietet eine Hochgeschwindigkeits-Technologie für die chemische Umwandlung dünner Oberflächen-Schichten von Plasten, Lacken, Überzügen, Druckfarben u. a. In einer Übersicht werden die Hauptrichtungen der angewandten Techniken, Apparate und Materialien dargestellt und der gegenwärtige Stand der Chemie und Technologie von elektronenstrahlbehandelten Polymersystemen diskutiert. Die vorgestellten Beispiele sind der Plast- und Papierbearbeitung entlehnt, ferner aus der Holzindistrie und aus anderen Industriezweigen.

Keywords

coatings; electron beams; fabrication; polymers; radiation effects; varnishes

Introduction

Low-energy electron beam curing, applying electron energies mainly between 150 and 350 keV, represents the most rapidly growing field of industrial radiation processing in the '80-es. This is demonstrated not only by the recent literature data and conferences focussing on radiation technologies — such as the series of International Meetings on Radiation Processing, or RADCURE Conferences [1, 2], — but also by the growing number of low energy EB machines installed in profit-oriented industry branches in full expansion [3]. In the last decade more than 250 low-energy EB machines have been installed for curing of thin layers of macromolecular systems, films and coatings on plastics, wood, metal and paper [3].

The main advantages of low-energy EB technologies over the conventional curing of coatings are the followings:

- Use of solventless, 100% polymerizing binders of less hazard and environmental pollution.

* Presented at the 3rd International Meeting on Radiation Interaction, 24—28 Sept., 1984, Leipzig — GDR

[1]) Address: H-1950 Budapest, XIV, Hungaria Körut 114

- Chain reaction curing initiated at low temperature, assuring complete curing at an unusually high speed.
- Extremely low energy consumption because of the high conversion yield of electric power into accelerated electrons in these modern machines, and mainly because of the efficiency of the localized radiation-initiation as compared to the conventional thermal curing.

The major technical limitation of the low-energy EB curing is the limited penetration depth. Tipically, 300 keV electrons can treat a plastic layer of maximum 400 g/m^2 i.e. about 0·4 mm of thickness.

The growing success of the low-energy EB technologies is based on one side by the new generation of the in-line adaptable, "self-shielded" EB machines requiring no special shielding or separated room of the former industrial sources of ionizing radiation. On the other side, this success is supported by the broadening veriety of new synthetic chemicals used, on basis of unsaturated, often acrylic-modified oligomers and multifunctional monomers, mainly acrylates.

Chemistry of EB-curable Polymeric Systems

The chemistry of EB-curable materials is making use of the long experiences of the UV-curing of coatings, laquers, printing inks, etc. In fact, all the photochemically curable coatings can be used also in EB technique even without photosentitizers. This means an easier storage and longer shelf-life for EB-curable materials as compared to UV-curable ones.

The unsaturated oligomer components of the EB coatings are mainly oligoesters and their acrylates, polyether acrylates, as well as epoxy acrylate resins.

Oligoester acrylates

Unsaturated oligoesters dissolved in vinyl monomers can be polymerized effectively by EB curing. Their reactivity however, can be greatly enhanced if modified by acrylic end-groups. Saturated and unsaturated oligoester acrylates offer excellent coating properties. A typical representative of this group:

$$CH_2 = CH—\underset{\displaystyle O}{\overset{\|}{C}}—O—(CH_2)_6—\left[—O—\underset{\displaystyle O}{\overset{\|}{C}}—(CH_2)_4—\underset{\displaystyle O}{\overset{\|}{C}}—O—(CH_2)_6—\right]_n—O—\underset{\displaystyle O}{\overset{\|}{C}}—CH = CH_2$$

Oligoester acrylates can be produced in two ways:
firstly, the polycondensated skeleton can be manufactured to which acrylic end-groups can be attached;
secondly, in case of simple oligoester acrylates, the reaction of (e.g.) dicarboxylic acid, pentaeritritol and acrylic acid can proceed in a one step process.

Polyether acrylates

These are producing soft, elastic coatings if cured by EB on flexible substrates. A typical example for such materials is the poly(propylene glycol)acrylate:

$$CH_2 = CH—\underset{\displaystyle O}{\overset{\|}{C}}—O—\left[—CH_2—\underset{\displaystyle CH_3}{\overset{}{CH}}—O—\right]_n—\underset{\displaystyle O}{\overset{\|}{C}}—CH = CH_2$$

Functionality of these polyether main-chain is usually between 2 and 6, so the activity of their acrylate derivatives can be varied between broad limits.

Urethane acrylates

Urethane based laquers, varnishes are proven in many practical applications. Actually different urethane acrylates are offered specially for EB curing. An example for that group:

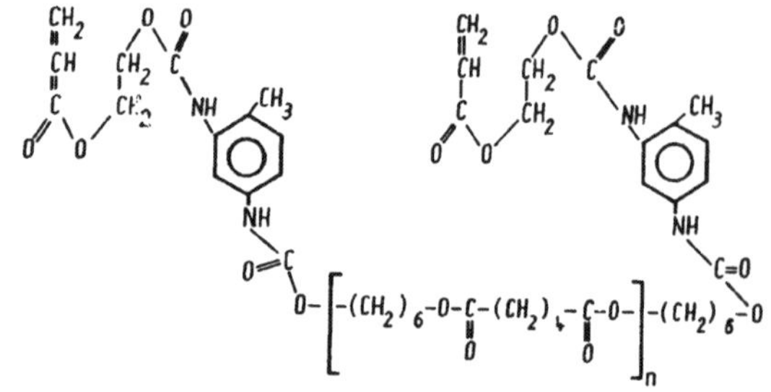

As it is seen, urethane-modified polyester main-chains are coupled with acrylic end-groups here.

Together with multifunctional monomers, a series of coating systems can be prepared with different performances and properties.

Epoxy acrylates

Epoxy resins are well known on their excellent binder properties but they have not been involved till recently in radiation curing. Here again, the acrylate derivatives are EB curable:

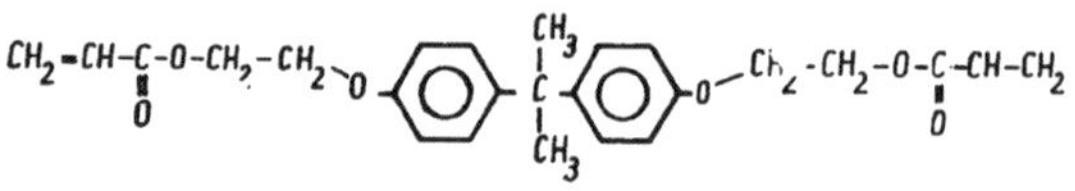

Repeating diane-expoxy chains before coupling with acrylates, coatings of different grade can be produced for metal, wood, paper or other substrates. Using methacrylic acid instead of acrylic one, vinylester resins are obtained, known from the excellent chemical resistance of their crosslinked polymers.

The EB radiation crosslinking of the above enumerated oligomers or prepolymers is performed in presence of a series of multifunctional acrylic monomers. Actually, several dozens of major chemical companies offer their different monomers and oligomers as well as coating mixtures formulated especially for EB curing [4].

EB-reactive silicones

Silicon-coated papers are used for pressure sensitive labels as well as in manufacturing of synthetic leather as "release paper". Several new silicon derivatives were developed and offered commercially for EB curing [5]:

$$—\overset{|}{\underset{|}{Si}}—(CH_2)_3—O—\underset{\displaystyle O}{\overset{\|}{C}}—CH = CH_2 \qquad —\overset{|}{\underset{|}{Si}}—(CH_2)_4—O—\underset{\displaystyle O}{\overset{\|}{C}}—CR' = R''$$

$$—\overset{|}{\underset{|}{Si}}—R—S—\underset{\displaystyle O}{\overset{\|}{C}}—CH = CH_2$$

These silicon acrylates (or similar vinyl silanes) can be reacted under EB radiation not only between each other but also with mercapto-silanes:

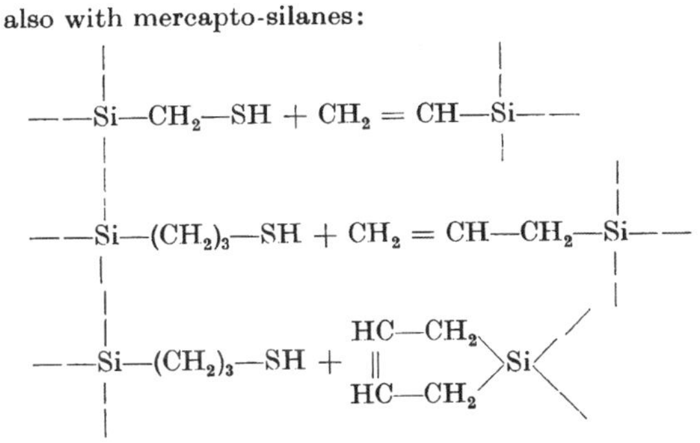

EB-reactive pressure-sensitive adhesives

The earlier technology of applying pressure sensitive (PS) adhesive layers on paper or plastic films can also be changed for EB curing [6]. Here again, the solventless, energy-saving technique of EB curing offers great technological advantages. EB curable PS adhesives are based mainly on acrylates e.g. polymer of butylacrylates containing some acrylic acid. Some other PS adhesives contain poly(vinyl alkyl ethers) in low quantities. Copolymers of vinylacetate and 2-hexylacrylate are excellent EB curable PS adhesives.

The EB technology of applying PS adhesives is very advantageous as compared to the hot-melt application, especially if heat-sensitive substrate (plastic base-film) is concerned.

Let us now survey briefly the teehnics of low-energy EB curing.

Low-energy EB Machines and Techniques

EB machines up to about 350 keV energy can easily be adapted to conveyor-type processing lines, treating webs of paper, plastics, textiles etc. Only light shielding is necessary to protect workers in the surroundings: the most advanced LE-EB machines are "self-shielded".

The IRELEC S type (French) machines apply scanned electron beams which can be adapted to conveyor width from 0·3 to 1·8 m [7]. Some typical data of these and the following machines are compared on Tab. 1.

Similarly scanned electron beam is used in the ESH-150 machine (FRG). The effective depth of penetration being 120 g/m² at 150 kV indicates an advanced technical solution for the beam window [8].

Electron accelerators with local shielding for industrial processing are offered by the Efremov Scientific Research Institute of Electrophysical Apparatuses (N.I.I.E.F.A.) in Leningrad, USSR. Beam voltage of these AURORA machines is between 300 and 500 kV, their scanning width (window) is between 0·5 and 2 m, and their beam power (P) can reach 25 kW [9].

Another family of low-energy EB machines is fabricated at the Institute of Nuclear Physics of the Sibirian Division of the Academy of Sciences, USSR, Novosibirsk. The accelerators ELV-3 are offered from E = 0·4 to 1 MeV and P = up to 20 kW, but they are typically not self-shielded [9].

In Japan the NISSIN High Voltage is producing low-energy EB units of scanned beam type, on the basis of the earlier High Voltage types [10].

The most advanced LE-EB machines are produced by CHARMILLES, ESI, Geneva, CH, under the licence of ENERGY SCIENCES Inc., Woburn, MA, USA [11]. These ELECTROCURTAIN® machines are linear cathode type selfshielded accelerators of very compact construction. A reduced-scale, laboratory unit has also been developed with beam densities comparable to the large-scale (industrial) machines. That means, with a reduced beam-width of 15 cm, one can try out industrial EB treatment processes with conveyor-speed practically equal to that of the industrial EB units of up to 2 m beam-width. A typical curing rate (conveyor-speed) of these ELECTROCURTAIN machines at 1 Mrd (10 kGy) dose delivery is 200 m/min. However, their most advanced types can deliver 1 Mrd for web of 1 m width, running with a speed up to 1350 m/min. All together more than 200 ELECTROCURTAIN units are working worldwide, according to recent informations [3].

Another low-energy EB processor is made by RPCI Co., Hayward, CA, USA. It is a multifilament, selfshielded EB machine, oriented longitudinally the electron guns along the direction of motion of the film-type product, typically magnetic tape [3].

Performance data of different low-energy EB machines are compared in Tab. 1. Not only their dose delivery but also their very efficient energy-conversion are remarkable: Several of them convert the input (total) electrical energy

Tab. 1. Comparison of some current low-energy EB machines

Type	Voltage kV	Beam Current mA	Beam Power kW	Scan width m	Effective* penetration g/m²	Dose delivery Mrd · m²/min	Dose delivery Mrd · g/s	Dose delivery kW	Energy Conversion Beam efficiency %	Energy Conversion Main efficiency %
IRELEC 200/100 (F)	200	100	20·0	0·3 – 1·8	60	400	400	4·0	20	16
IRELEC 300/75 (F)	300	75	22·5	0·3 – 1·8	280	300	1400	14·0	63	50
ESH 150 (FRG)	150	100	15·0	1·0 – 1·3	120	180	360	3·6	24	
AURORA II. (SU)	300 – 500		25·0	0·5 – 2·0						
NISSIN HV (JAP)	300	25	7·5	0·45	350	50	290	2·9	38	30
	300	65	19·5	1·2	350	130	760	7·6	39	30
	300	100	30·0	1·8	350	200	1170	11·7	39	30
ELECTROCURTAIN (USA)										
LAB UNIT	175	10	1·75	0·15	170	34	97	0·97	54	
CB 175/90/60	175	60	10·5	0·9	170	180	285	2·85	48	
SIMPLE FILAMENT UNIT (GENEVA UNIT)	150 – 250			0·9	110 – 350	270			up to 70	
TRIPLE FILAMENT UNIT	150 – 300			0·5 – 2·0	110 – 400	1350			up to 70	

*) given by the manufacturers, depending on window thickness

into electron-energy absorbed in the target, with an efficiency higher than 50%. The beam efficiency — not counting energy consumption of the service machines — is even higher. Considering the excellently focussing electron-optics of these machines, as well as the practically total electron-absorption in the working piece, these efficiencies are never seen before in the radiation technologies.

Low energy EB curing in plastics processing

The far largest volume of industrial electron processing today is the radiation crosslinking of different polyolefins in form of films, hoses, tubings ets., to produce heat-shrinkable materials for packaging and mainly for the electrical industry. The electron sources used here till recently were of high energy, usually of several MeV. The limited penetration of low energy (up to 300 keV) EB has been considered formerly as insufficient for that particular purpose of crosslinking.

In our experiences we compared low-energy (175 keV) EB treatment with conventional method of crosslinking (by 1,8 MeV EB) in a practical application, producing heat-shrinkable polyolefin packaging films of 0·200 mm thickness, made by the following two alternative ways (A and B).

Electron sources	A 2 MeV *Van de Graaff* VEB TuR, Dresden	B ELECTROCURTAIN CB 175 Lab unit Energy Sci,. Inc. Woburn,Mass.
Voltage range, kV	1200 – 2000	150 – 175
Beam current, max., mA	0·20	10·0
Beam power, max., W	400	1750

Material (in both cases): Polyethylene (ALKATHENE 501)
- extruded hose of 0·18 – 0·20 mm wall thickness
- flattened width (of double wall) 103 mm
- unit mass: 33 – 36 g/m

Actual conditions of the EB treatment:

Passage under the beam:	A Single passage in 5 folding	B One passage from one face, then a second passage from the other face
Irradiated layers	5 × 2 × 0·2 mm³	0·5 · 2 · 0·2 mm³
Transport speed, m/min	1·44	15·0
Production yield, kg/h	2·88	15·0
Dose delivery, kGy	40·0	40·0
Actual voltage, kV	1800	175
Beam current, μA	180	2900
Beam power, W	324	507
PE-absorbed beam power, W	32	150
Beam utilization, %	9·9	30
Total power input, kW	8·0	3·5
Total power efficiency, %	0·4	4·3

The polyethylene films crosslinked by these two alternative ways were equally adaptable under the same conditions to the stretching machine were heat-shrinkable packaging materials have been produced. All the performances and properties of the final products where practically the same.

On the basis of our experiences, we consider that there are many similar applications of crosslinked polyolefin products, where low-energy EB curing is applicable, even if a certain inhomogeneity is obtained in the crosslink-density in function of the depth. Even a surface-crosslinking can be sufficiently effective for many (partially) heat-resistant or heat-shrinkable products.

Wood-panel coating by low-energy EB curing

The following part is a short summary of a case-study where a 200 keV EB machine is used in industrial-scale production line of coated wood panels of 1·2 m × 2·5 m typical dimensions [12].

Electron source applied: ELECTROCURTAIN CB 200/140/90

Voltage: 200 kV, beam width: 140 cm, beam current: 90 mA

Beam power: 18 kW

Dose distribution in the depth of different layers is shown in Fig. 1.

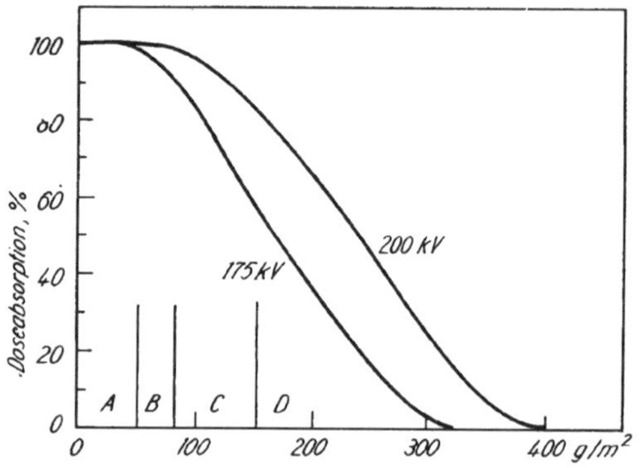

Fig. 1. Dose distribution as a function of thickness

Layer A — coating 50 g/m²; B — decor-paper 30 g/m²; C — adhesive 70 g/m²; D — wood panel 20 mm

Total dose (for the coating): 50 kGy at conveyor speed of 52 m/min

As it is seen in the Figure, EB cures in one step two different polymerizable layers, below and above the printed decorative paper or vinyl sheet.

To compare the energy consumption of the above detailed wood-panel coating line of low-energy EB process, we have considered two conventional curing line working actually in our country. A modern, semi-automatic, window-processing line was installed in Hungary in 1984 — the major part of machinery being delivered by O. Dürr, FRG. The line is producing yearly 250000 m² of complete, lacquered, ready-to-install window. The conventional heat curing of the lacquers occurs in a room of 18 × 18 × 2·5 m³ equipped with a continuously moving conveyor and a heating power of nearly 2 million of kcal/h (= more than 2300 kW). The conveyor speed is around 1 m/min. Another example to compare the energy consumption of EB curing is another in Hungary producing yearly 250000 m² of pre-assembled parquet-panels (in units of 137 × 3000 mm²), varnished in a conventional heat tunel. This heat-tunel is of 18 m long, consuming about 300 kW heat-energy, working at a conveyor speed of 4 m/min.

The specific energy consumption for curing 1 m² of lacquered surface could be by 2 or 3 orders of magnitude less in both examples, if cured by low-energy EB.

Coating of paper and metal by low-energy EB

Coating of paper, forming overprint varnishes, release-coating, pressure-sensitive coating is one of the most promising field of low-energy EB curing where the extremely high producing (web-) speed, reaching several hundred meters/min, is an important and very attractive advantage. Supermooth coatings, buried layer coatings (between sandwiches) can be produced by EB. Vaccum-metallized layer containing coated multilayer systems are processed actually on several EB lines around the world [3, 4].

Coating of metal sheets, using UV curable printing inks and varnishes is a known method. In Japan precoated steel sheets are produced by low-energy EB coating, having not only good appearance, hardness and gloss, but also excellent post-formability [13].

Curing of binders in magnetic media

By the growing consumption of tape-recorders, video cassettes as well as professional and personal computers, there is a growing need in magnetic media (magnetic tapes and floppy disks). Curing of low viscosity — mainly urethane based — binders containing metal oxides on a polymer film substrate can be well done by low-energy EB curing. Because of the heat-sensitivity of the substrate, low temperature fast curing is desirable. According to a recent report [3], more than 80% of the tape manufactures in the U.S., Japan and Western Europe have electron curing equipment. More than 20 small-scale (laboratory and pilot scale) EB machines are working on the development of magnetic media, and at lest 6 full-scale EB processing lines are producing them [3]. EB-sensitive binders are based mainly on polyurethane oligomers, typical curing dose is 5 Mrd, at a curing speed of 150 m/min.

Other application of low-energy EB curing: microelectronics

The limited space of present lecture doesn't allow to discuss in details the highly interesting field of EB curable resists applied in the fabrication of silicon-based integrated circuits. Positive and negative EB-resists have already been introduced in the manufacturing of VLSI (very large scale integration) integrated circuits, achiebing a line-resolution of better than 1 micron. The EB-curable polymeric systems form a class of speciality chemicals used in microelectronics [14].

The new types of EB curable resists are making use of the former experiences on radiation grafting, too [15].

Received November 28, 1984

References

[1] a) 4th International Meeting on Radiation Processing, Dubrovnik, 4—8 Oct. 1982. See: Radiat. Phys. Chem. **22** (1983) No 1—5
b) 5th International Meeting on Radiation Processing, San Diego, California, 22—26 Oct. 1984

[2] a) RADCURE '83, Lausanne, 9—11 May 1983. See: Coating **16** (1983) No 6—8
b) RADCURE '84, Atlanta, 1—13 Sept. 1984

[3] *S. V. Nablo*, The status of the industrial use of selfshielded electron processors. Lecture at the Int. Conf. on Appl. of Radioisotopes and Radiation in Industrial Development, 1—3 March, 1984, Bombay

[4] *T. Czvikovszky. K. Helyes, Gy. Madai*, Papiripar (Paper Industry, in Hungarian) 27 (1983) 197

[5] *D. R. Thomas, J. D. Jones*, Siliconising with radiation curing system. RADCURE '83 Conf., Lausanne, 9—11 May 1983

[6] *K. C. Stueben*, Adhesives Age **20** (1977) No. 5, 16

[7] Machine Prospectus of IRELEC, Ass. Ind. Dauphinoise, F-38240 Meylan, France

[8] Machine Prospectus of ESH-150, 0. Dürr Anlagenbau GmbH, D-7000 Stuttgart 40, FRG

[9] Machine Prospectus of TECHSNABEXPORT, 121200 Moscow, USSR

[10] Machine Prospectus of NISSIN High Voltage, Japan

[11] Machine Prospectus of ELECTROCURTAIN, Charmilles S.A., Energy Sci. Int., CH-1211, Genève 13, Suisse

[12] *D. French, B. S. Quintal, S. V. Nablo*, Adhäsion **83** (1983) 152

[13] *S. Fujioka, J. Fujikawa*, Radiat. Phys. Chem. 18 (1981) 865

[14] *H. J. Lorkowski*, Plaste u. Kautschuk **30** (1983) 421

[15] *M. Morita* et al., J. Vac. Sci. Technol. B 1/4 (1983) 1171

Isotopenpraxis **21** (1985) 11, S. 383

Buchbesprechung

Evaluation of Occupational and Environmental Exposures to Radon and Radon Daughters in the United States

Bethesda, Mai 1984, 204 S., 12 Abb., 43 Tab. (NCRP Report No. 78); kart. $ 15.—

Band 78 des "National Council on Radiation Protection and Measurement" der USA berichtet ausführlich über die berufliche Strahlenbelastung von Bergarbeitern und die natürliche Strahlenbelastung der Bevölkerung durch die Inhalation von Radon und seinen Folgeprodukten. Radon und insbesondere die an Aerosole angelagerten kurzlebigen Folgeprodukte führen im tracheobronchialen System und pulmonalen Bereich durch die Emission energiereicher Alphastrahlung zu einer erheblichen Strahlenbelastung der Epithelzellen und stehen in unmittelbarem Zusammenhang mit der Manifestation von Bronchialkarzinomen bei Bergarbeitern. Darüber hinaus beeinflussen verschiedene chemische Karzinogene in den Gruben und die Rauchgewohnheiten der Bergarbeiter das Bronchialkrebsrisiko nicht unerheblich.

Auf der Grundlage eines umfangreichen Untersuchungsmaterials aus dem Uraniumbergbau der USA, unter Einbeziehung von Erfahrungen aus Schweden, Kanada und der ČSSR, werden tiefere Einblicke in die Zusammenhänge zwischen Strahlenbelastung und Erkrankungshäufigkeit an Bronchialkarzinomen vermittelt. Unter den zahlreichen Einzelproblemen nehmen die Charakterisierung von Radonquellen, die Dosimetrie von Radon und Radonfolgeprodukten im Bergbau, die Festlegung von Dosis-Konversionsfaktoren, das Konzept der potentiellen Alphaenergie-Konzentration (working level) und die Empfehlung von Dosisgrenzwerten breiten Raum ein. Die Auswertung epidemiologischer Daten über das Bronchialkrebsrisiko im Uraniumbergbau sowie radiobiologische Beobachtungen an Versuchstieren gestatten Abschätzungen des Risikos für die beruflich nicht strahlenexponierte Bevölkerung.

Ein Literaturverzeichnis mit 294 Titeln, in dem einschlägige Publikationen bis Anfang der achtziger Jahre Berücksichtigung finden, vervollständigt den Band. Bei der Wiedergabe deutschsprachiger Namen und Begriffe sind jedoch viele Druckfehler unterlaufen.

Für alle unmittelbar auf dem Gebiet des Strahlenschutzes tätigen Wissenschaftler, aber auch für Onkologen, Epidemiologen und Arbeitsmediziner ist die vorliegende Publikation von großem Aussagewert.

W. Stolz

Isotopenpraxis 21 (1985) 11, S. 384 bis 387

Further Studies on Radiation Dosimetry of Some Amino-Anthraquinone Dyes

M. M. Abou Sekkina[1]), *S. S. Assar*

(Chemistry Department, Faculty of Science, Tanta University, Tanta, Egypt)

Eight aminoanthraquinone dyes were prepared, recrystallized and identified. The prepared materials were investigated by means of infrared absorption spectra. Then the materials were subjected to different doses of γ-radiation up to 4·5 × 10⁶ rd. The infrared absorption spectra of each material were measured after each radiation dose and after standing for two months out of the radiation field. Results obtained indicated the existence of Zwitter ion structure for 1-amino-anthraquinone-2-sulfonic acid and a sequence was established for the relative stability of the different dyes against γ-irradiation. Calibration curves were constructed indicating that most of the investigated dyes are capable to be used as dosimeters up to 4·5 × 10⁶ rd. A correlation between the extent of γ-radiation damage and the energy of hydrogen bonding was evaluated. Finally, a detailed mechanism was put forward for the interaction of ionizing radiation with the investigated dyes.

Acht Aminoanthrachinon-Farbstoffe wurden synthetisiert, rekristallisiert und identifiziert. Es wurden IR-Spektren aufgenommen. Die Farbstoffe wurden einer γ-Bestrahlung bis zu 4·5 × 10⁶ rd ausgesetzt und die IR-Spektren vor und nach der Bestrahlung verglichen, wobei die bestrahlten Proben 2 Monate lagerten. Die Ergebnisse deuten im Falle der 1-Amino-anthrachinon-2-sulfonsäure auf Zwitterionenstruktur hin. Es konnte eine Rangfolge der Stabilität der Farbstoffe gegenüber der γ-Strahlung aufgestellt werden. Durch Eichkurven konnte gezeigt werden, daß die meisten der Farbstoffe als Dosimeter bis zu 4·5 × 10⁶ rd verwendet werden können. Eine Korrelation zwischen der eingetretenen Zerstörung und der Energie der Wasserstoffbindungen konnte aufgestellt werden. Ein detaillierter Mechanismus für die Wechselwirkung der ionisierenden Strahlung und den untersuchten Farbstoffen wurde vorgeschlagen.

Keywords

anthraquinones; dosimetry; dyes; gamma radiation; infrared spectra; radiation doses; radiation effects

Introduction

The detailed mechanism of γ-radiolysis is not known at present, but the basic process are known. They include the ionization and excitation of molecules followed by dissociation with the formation of free radicals or atoms [1]. Thus, *Harrington* [2], studying the damaging effect of radiation on plastics and elastomers; the effect of gamma radiation exposures of up to $1{\cdot}5 \times 10^8$ rd on the mechanical properties were listed for 48 materials.

The effect of γ-radiation exposure on the spectral properties of various materials have been investigated [3, 4]. *Singel* and *Whitten* [5] worked on the effect of γ-radiation on the life time of anthracene. They deduced that γ-radiation produces traps with trapping cross-section of a unit cell as a result of damaging. Irradiation of benzene has been reported [6] to yield a liquid polymer. *Shaw* [7] in studying the effect of γ-irradiation on iodobenzene deduced the formation of diphenyliodonium as well as of iodine in pile irradiation. *Porroek* et al. [8] irradiated a mixture of iodobenzene with diethylamine, also they investigated the irradiation of o- and p-iodoanisole and show that the electron releasing methyl group might favour heterolysis rather than homolysis of C-I bond.

Arylation of aromatic compounds by radicals generated by γ-radiolysis of bromobenzene have been shown by *Everard* and *Swan* [9]. Extensive investigations on the action of ionizing radiation on apliphatic ethers have been carried out by *Newton* [10] and by *Bakh* et al. [11] who measured the yields of carbonyl compounds, alcohols and peroxides; γ-radiolysis of allylphenyl ether yield phenol and a polymer as the main products. The effect of -irradiation on the infrared absorption spectra of natural wool fibers was previously investigated by *Moharram* et al. [12], showing that the predominant effect of γ-radiation on wool is the oxidation of systine. The increase of absorbance with the increase of γ-doses indicates a decrease in the degree of crystallinity.

Hanafi et al. [13] investigated the effect of γ-radiation on the electrical conductivity and infrared absorption spectra of indium oxide showing that irradiation leads to predominantly donor type defects.

Elagramy et al. [14] studied the induced electrical conductivity of gamma irradiated 8-hydroxyquinoline metal complexes deducing an increase in activation energy for conduction and a decrease in the obtained value of electrical conductivity as a result of γ-irradiation.

Recently, Abou Sekkina and *Shakra* [15] investigated the effect of γ-irradiation on the electrical and optical properties of solid ozodyes. The relative stability of the different azodyes had been evaluated and a correlation was put forward with the type of substitution in the organic molecule.

In this respect, the ir spectra of eight related amino-anthraquinone dyes before and after γ-radiation doses were studied. These were carried out with an attempt of evaluating a detailed account of the various changes that probably existed as a result of γ-irradiation and the relative stability of the different dyes in correlation with the type of substitution and the strength of hydrogen bonding.

[1]) Author for correspondence to be addressed

Experimental

a) *Preparation of aminoanthraquinone (A.A.Q.) dyes.* The present investigated aminoanthraquinone dyes 1-amino-anthraquinone (I); 2-amino-anthraquinone (II); 1,8-diamino-anthraquinone (III); 1-amino-anthraquinone-2-sulfonic acid (IV); 1-amino-4-hydroxy-anthraquinone (V); 1-amino-2-bromo-hydroxy-anthraquinone (VI); 1-amino-anthraquinone-2-carboxylic acid (VII); 1-amino-2-aldehydeanthraquinone (VIII) were prepared and recrystallized [16, 17].

b) *Gamma radiation exposure.* The present freshly prepared group of related aminoanthraquinone dyes (I-VIII) were exposed, to various γ-radiation doses using (^{60}Co gamma cell (250 Ci); The radiation energy was taken at a dose of (260 rd/min) at room temperature. The investigated samples were exposed to doses of $0{\cdot}6 \times 10^6 - 4{\cdot}5 \times 10^6$ rd.

c) *Infrared absorption spectra measurements.* The various (A.A.Q.) dyes in concentration of 1:350 of KBr, spec. pure (SP) were ground in an agate mortar. The gound mixes were pressed into transparent pellets (220 kg/cm^2) immediately using a suitable die which could be evacuated.

The percentage transmittance was automatically registered against wavenumber using a "PYE UNICAM SP 1000" Spectrophotometer (L.T.D. Cambridge England) in the range of "3800 – 625 cm^{-1}. The spectra were recorded before and soonly after each radiation dose. Moreover, the spectra were also recorded after removing the specimens from radiation, field for a sufficient time (two months) to detect any liable reversible changes if present.

Results and discussion

In the present investigation, the infrared absorption spectra of the dyes (I – VIII) before and after irradiation doses up to $4{\cdot}5 \times 10^6$ rd, are investigated.

Furthermore, in order to detect any liable reversible change if present in the effect of γ-irradiation, the spectra were further taken once more again after standing for two months from the final radiation dose.

From the results obtained one can summarize the effect of γ-irradiation doses on the infrared absorption spectra of the investigated aminoanthraquinone dyes as follows:

a) The absorption bands are gradually shifted towards high wavenumbers corresponding to higher energies.This is probably correlated to the formation of decomposition centres and/or polarization centres vibrating at higher energies.

b) The intensities of some absorption bands are gradually decreased as a function of radiation doses. This observation is correlated with the induced negative absorption which propagated in the same direction.

c) Some bands disappear and others are created as a result of γ-radiolysis. Moreover, there are a noticeable changes towards broadening and weakening of the absorption bands and these are probably due to the decreased degree of crystallinity [12].

However, in irradiation with γ-rays the compton effect [18] has the largest existance causing atomic displacements within the material. A case which takes place with material composed of low atomic numbers [19]. Thus, irradiation with these rays constitutes atomic displacements by the compton electrons and/or free radicals formation and recombination [9]. This inturn may produce traps with trapping cross section of a unit cell as a result of gamma radiation damaging [5]. This was also in accordance with the previous foundation of *Eisen* [20] that damage produced by irradiation with high energy level of γ-doses is expected to consist mainly of randumly located interstitial vacancy spacings which depends on the energy level of γ-irradiation. In order to detect if any liable reversible changes for radiation effect, the infrared absorption spectra of the same specimens and at the same conditions are recorded after standing out of the radiation field for two months from the final radiation dose. The obtained results indicate that the occurred changes are irreversible for the dyes, I, II, III, IV and VIII which suffer from strong effects of γ-irradiation (gamma radiation damage). This could be attributed to the destructive action of γ-irradiation on these dyes which probably causes some type of chemical changes. On the other hand, the effect of γ-irradiation doses on dyes V, VI and VII, which show relatively less pronounced effects, have a reversible character. This was deduced from measurements of their infrared absorption spectra after two months showing a great similarity with those measured before irradiation. Besides, little changes are still remained in the obtained spectra after time. This may be explainable on the basis that, over the induction period, the point defects are annealed out or annihilate [20 – 22] and the energy released at the site of recombination results in the formation of decomposition nuclei [23] in addition to those induced upon irradiation consequently, leaving some little permanent effect.

The absorption bands located in the region above 3000 cm^{-1}, which are assigned to be due to stretching vibrations [24] of aromatic C—H, O—H and NH_2 groups are completely destroyed for all investigated dyes except the dye IV. On the other hand, the intensities of the bands in the range 1685 – 1260 cm^{-1} which correspond to $\nu_{C=O}$, $\nu_{C=C}$ and ν_{C-C} of quinonoid structure for the dyes I, II, III, V and VIII are decreased with increasing the doses until they are completely disappeared there.

This indicates the degradation of these dyes by γ-irradiation. It is of interest to note that for the dye VIII, which is the aldhydic derivative of aminoanthraquinone, a new band at 1690 cm^{-1} is observed at the first irradiation dose. This is probably correlated to the partial oxidation of CHO group to-COOH group as a result of the oxidizing nature of γ-irradiation. This was in accordance with that previously deduced by *Moharram* [12] et al. This band was disappeared at the second irradiation dose. This may be further explained on the basis that it undergoes decarboxylation which is a chemical change having a permanent character as being due to the destructive action of γ-radiolysis. The ir spectra of the dye VII show a band at 3145 cm^{-1} corresponding to streching vibrations of the – OH group of the carboxylic group. This band does not appear after irradiation of the sample. This is a further prove that the dye VII undergoes decarboxylation which involves chemical changes having a permanent character as being due to the destructive action of γ-radiolysis.

It is attractive point of view to note that the infrared absorption spectra of the dye IV which is relatively, more affected by γ-irradiation, exhibits absorption bands at 2640 and 3040 cm^{-1} which has been attributed to ν_{HN_3} [25] indicating that it has Zwutter ion structure. The intensity of the two bands decreases with increasing γ-radiation doses until disappeared there at 4·5 Mrd. Moreover, two bands located at 3280 and 3460 cm^{-1}, attributed to ν_{sym} NH_2 and ν_{asym} NH_2 appeared after first irradiation dose. On the other hand, the bands corresponding to δ aromatic ν_{C-H}, ν_{C-C} or quinone and stretching vibrations of sulphonic acid group become sharp and definite. Thus, one can conclude that the γ-irradiation prevents Zwitter ion formation, also, this dye has low stability towards the high ionizing γ-radiation.

Comparing the magnitude of γ-radiation damage on the various investigated dyes one can arrange them according to their relative stability towards the effect of high ionizing γ-radiation doses as follows:

$$\text{III} < \text{I} < \text{II} < \text{V} < \text{VIII} \lessapprox \text{VI} < \text{VII} < \text{IV}\,.$$

This means that: (i) dye III is the least one affected by γ-irradiation and hence it is the most stable dye. (ii)-dye IV is the highest one affected by γ-irradiation and hence it is the lowest stable one.

It had been seen from earlier work, made by the same author [24], that there was a conformity between the sequence of the energies of hydrogen bonding and that of relative stability of the different dyes towards the effect of γ-irradiation. Thus a trend is deduced that the more stronger the hydrogen bonding the greater stability of the investigated dye towards the effect of γ-irradiation.

From results obtained, it was noted that the absorption of some bands decreases linearly with increasing the energy of γ-radiation as being correlated with the decreased degree of crystallinity [12].

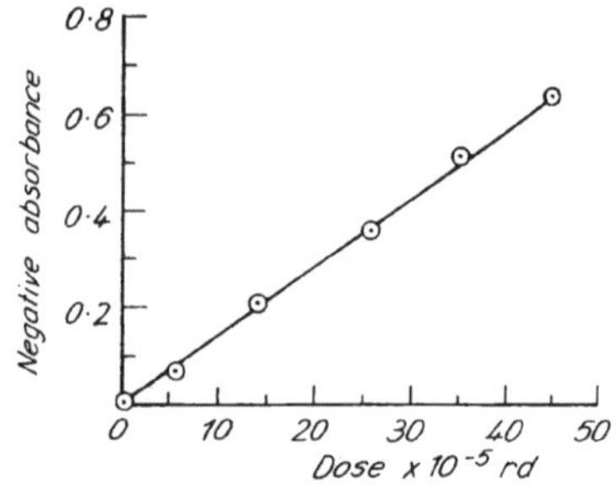

Fig. 1. Correlation of the negative absorbance with γ-radiation doses for the dye I (ν = 709 cm^{-1})

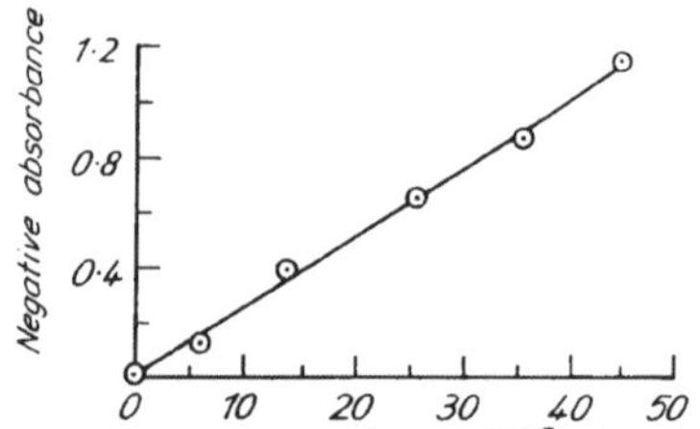

Fig. 2. Correlation of the negative absorbance with γ-radiation doses for the dye VI (ν = 1590 cm^{-1})

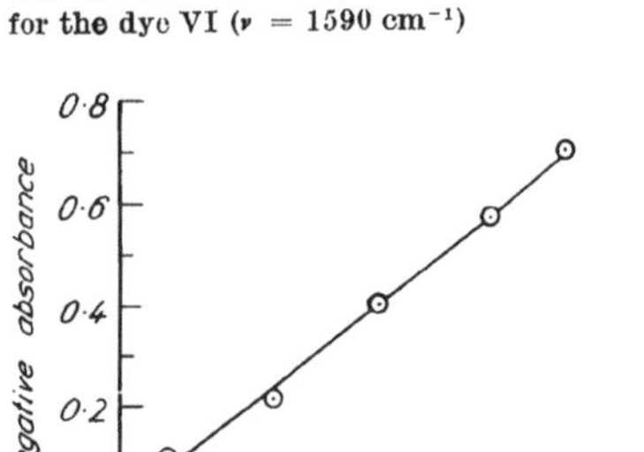

Fig. 3. Correlation of the negative absorbance with γ-radiation doses for the dye VII (ν = 900 cm^{-1})

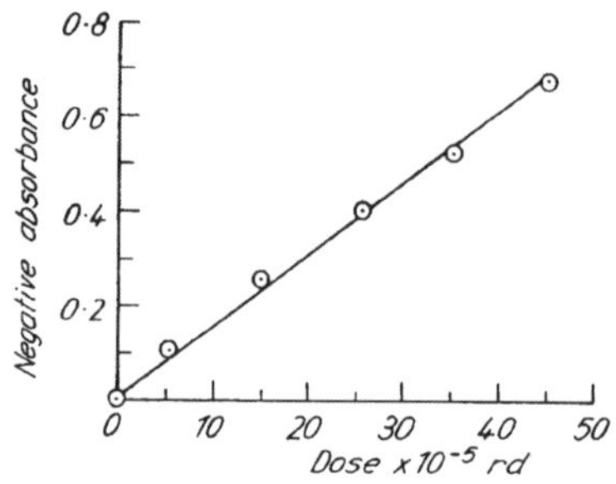

Fig. 4. Correlation of the negative absorbance with γ-radiation doses for the dye XII (ν = 955 cm^{-1})

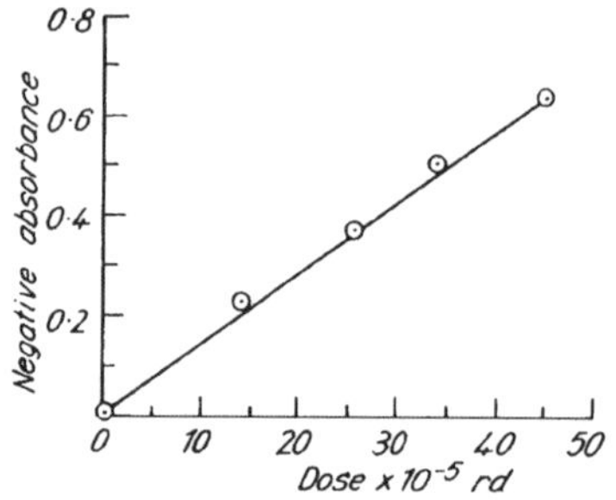

Fig. 5. Correlation of the negative absorbance with γ-radiation doses for the dye XIII (ν = 960 cm^{-1})

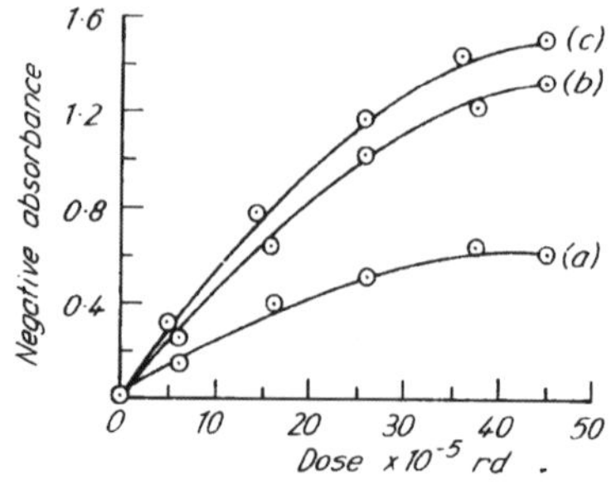

Fig. 6. Correlation of the negative absorbance with γ-radiation doses for (a) — the dye II (ν = 848 cm^{-1}); (b) — the dye IV (ν = 795 cm^{-1}); (c) — the dye V (ν = 1049 cm^{-1})

.Thus, this indicates the possibility of using these dyes as dosimeter in the range of doses up to 4·5 Mrd. Thus, the correlation between the negative absorbance of the characteristic bands and energy of γ-radiation doses are successfully used as a good calibration curve for dosimetry taking into account the bands at 709, 1590, 900, 955 and 960 cm^{-1} for the dyes I, V, VI, VII and VIII respectively (Fig. 1—5). For the other dyes, II, III, IV the relationship between the negative absorbance and energy of radiation doses are not straight lines for any band in the range investigated (see Fig. 6). indicating the impossibility of using these dyes as a dosimeters.

Suggested mechanism for the effect of γ-radiation doses on aminoanthraquinone dyes

From the above mentioned results and discussion, it may be possible to suggest a mechanism for the effect of γ-radiation on the investigated aminoanthraquinone dyes as follows:

1) The first step is the ionization of the dye giving rise to a positively charged aminoanthraquinone cation which has a sufficient stability due to resonance within the anthraquinone rings. At the same time, the free electron is captured by another molecule of aminoanthraquinone dye forming an anion which also has sufficient stability.

2) The second step is that, neutralization occurs by the reaction of the anion and cation formed displaying an excited molecule and another stable one.

3) The third step is a minor for reversible change or major (dissociation leading to free radical formation and/or destruction of the excited molecule to the observed amorphous or gummy products) for irreversible changes. During these steps lattice defects may occur. The created new lattice defects may probably lead to the observed negative infrared absorption. In certain cases, when these defects run linearly as a function of radiation doses, it makes the possibility of using these phenomena to construct a calibration curve to account the energies of γ-radiation doses, and hence one can conclude that most of these investigated dyes are capable to be used as dosimeters within the investigated range.

Conclusions

Accordingly, it was concluded that the high radiation resistance of dye IV is correlated with the existance of Zwitter ion structure. A sequence was established for the relative stability of the various aminoanthraquinone dyes towards γ-irradiation doses. There was a good agreement between this sequence and that construced for the relative energies of their hydrogen bonding denoting that the higher the energy of hydrogen bonding the greater would be their stability towards γ-irradiation. A detailed mechanism, for the interaction of ionizing radiation with the investigated aminoanthraquinone dyes was put forward.

Received February 25, 1985

References

[1] *Y. A. Gerasimov*, "Physical Chemistry" English Translation Mir Publishers, Vol. 2 (1974) 255
[2] *R. Harrington*, Nucleonics **14** (1956) No. 9,70
[3] *F. M. Ismail, E. M. Ibrahim*, Bull. N.C.R. **3** (1977) 27
[4] *A. Bishay, A. Kinawi*, Proc. Intern. Conf. "Physics of Non-crystalline Solids" North-Holand Pub. Co. Amsterdam 1965
[5] *H. Singel, W. B. Whitten*, J. Chem. Phys. **52** (1970) 1956
[6] *B. Batrick, B. Burton*, J. Am. Chem. Soc. **76** (1954) 2626
[7] *S. Shaw*, J. Am. Chem. Soc. **73** (1951) 443
[8] *J. D. Porroek, G. A. Swan, D. Wright*, J. Am. Chem. Soc. **84** (1962) 911
[9] *A. F. Everard, G. A. Swan*, J. Am. Chem. Soc. **84** (1962) 914
[10] *E. Newton*, J. Phys. Chem. **61** (1957) 1485
[11] *Bakh, Medvedarsky* and *Sarvaea*, Proc. 2ed. "United Nation Conference on the Peaceful uses of Atomic Energy Vol. XXIX, (1958) 128
[12] *M. A. Moharram, S. M. Rabie*, Isotope & Rad. Res. **10** (1978) 91
[13] *Z. M. Hanafi, E. M. H. Ibrahim, F. M. Ismail*, Egypt J. Phys. 8 (1977) 143
[14] *A. M. Elagramy, H. F. Aly, F. M. Abdel Kerim, M. H. Atya*, Isotopenpraxis **15** (1979) 223
[15] *M. M. Abou Sekkina, S. Shakara*, Pakistan Journal of Sientific, Research 32 (1980) 131
[16] *K. Bredereck, L. Banzhaf, E. Koch*, Chem. Ber. **105** (1972) 377
[17] *S. A. Metwally*, J. Appl. Chem. Biotechnol. **25** (1975) 161
[18] *R. D. Evans*, "The Atomic Nucleus" New York: Mc Graw-Hill 1967
[19] *G. J. Bienes, G. H. Vineyard*, "Radiation Effects in Solids" New York: Interscience 1957
[20] *F. H. Eisen*, Phys. Rev. **123** (1961) 736
[21] *M. J. Smith*, J. Appl. Phys. **4** (1963) 2878
[22] *M. J. Smith*, "Semiconductors", Cambridge: Univ. Press 1959
[23] *E. G. Prout, P. J. Herley*, Phys. Chem. **66** (1962) 961
[24] *M. M. Abou Sekkina, S. S. Assar*, J. Inorg. Nucl. Chem., in press
[25] *K. Nakanishi*, "Infrared Spectra and Structure of Organic Compounds" Moscow: Izd. Mir. 1965, p. 47

Isotopenpraxis **21** (1985) 11, S. 387 bis 391

Further Investigation on the Action of ^{60}Co-γ-Rays on the Optical Properties of Some Related Organic Ligands and Their Solid Transition Metal Complexes

M. M. Abou Sekkina[1]), *M. Y. El-Sheikh*

(Chemistry Department, Faculty of Science, Tanta University, Tanta, Egypt)

2,6-N,N'-diacetyldiaminopyridine (daapH), 2,6-N,N'-dibenzoyldiaminopyridine (dbapH) ligands and some of their solid transition metal complexes were purely prepared and recrystallized. The Nujol mull electronic absorption spectra have been undertaken for these materials before and after each γ-radiation doses of up to 10^9 rd. Results obtained indicated that the electronic absorption spectral bands of material investigated suffer shifts, changes in intensity, broadening, creation of new bands and/or completely disappearance of others as a result of radiation damage. A sequence of the relative resistance of materials investigated towards gamma radiation damage was put forward.

2,6-N,N'-Diacetyldiaminopyridin (daapH), 2,6-N,N'-Dibenzoyldiaminopyridin (dbapH) Liganden und einige ihrer Übergangsmetallkomplexe wurden hergestellt und durch Rekristallisation gereinigt. Die Nujol-Mull-Elektronen-absorptionsspektren von diesen Verbindungen wurden aufgenommen, und zwar vor und nach einer γ-Bestrahlung bis zu 10^9 rd. Die Elektronenabsorptionsspektren zeigten Bandenverschiebungen und -verbreiterungen, das Erscheinen neuer Banden und das Verschwinden von Banden als Folge der Bestrahlung. Eine Reihenfolge der relativen Widerstandsfähigkeit gegenüber γ-Strahlung konnte aufgestellt werden.

Keywords

cobalt 60; gamma radiation; ligands; transition metal complexes; ultraviolet spectra

Introduction

The interaction of ionizing radiations with some solids causes alteration in their spectral and sometimes chemical properties. The aim of many investigators has been to use the radiation induced changes of such properties as a means for measuring irradiation effect [1].

The series of fine structure bands in the 230–270 mμ region of the spectra of benzene and those of aromatic hydrocarbons are associated with vibrational effects on the Π-Π* transitions [2]. Thus, pyridine [2] have a series of

[1]) Author for correspondence to be addressed

fine structure absorption bands of lower intensity at 240 to 300 mμ. The spectra and resonance of -CONH group have been studied and discussed [3]. The absorption bands of this group appear within the range of 320 – 355 mμ. *Burger* [4] showed that the spectra of metal chelates are often similar to those of their corresponding organic ligands molecules. *Barbucci* et al. [5] found that the spectra of the transition metal complexes have no considerable change whether in the solid state or in solution. *Akaiwa* et al. [6] found that the ultraviolet absorption spectra frequencies of adducts decrease in the order: Cu(II) < Ni(II) > Co(II) > Zn(II). *Bajue* et al. [7] in studying the electronic spectra of some Co(III) complexes noticed the disappearance of the intense M-L charge transfer bands in the spectrum of each complex.

For the action of gamma radiation, *Harrington* [8] studied the damaging action of gamma radiation on elastomers and plastics. *Bishay* and *Kinawi* [9], *Ismail* and *Ibrahim* [10] studied the effects of gamma radiation exposure on the spectral properties of various materials.

As far as the authors are aware no mention had been cited in the literature data concerning the effect of this ionizing radiation on the ultraviolet absorption spectra of our prepared trensition metal complexes, which is the major goal of the present investigations.

Experimental

a) *Material Preparation*

This constitutes the preparation of two organic ligands and their different transition metal complexes. The two organic ligands are 2,6,N′,N′-diacetyldiaminopyridine (daapH) and 2,6,N,N′-dibenzoyldiaminopyridine(dbapH). These were prepared by the usual chemical procedure, recrystalized from water and benzene – alcohol mixture, respectively. The three transition metal complexes of $Co(daapH)_2Cl_2$; $Ni(daapH)_2Cl_2$ and $Ni(dbapH)_2Cl_2$ were prepared by the following procedure:

A solution of 0·20 mol of the prapared organic ligand in ethanol was added to the solution of the corresponding transition metal chloride (0·01 mol, in ethanol with stirring). The obtained crystalline precipitate was decanted, filtered off, washed with ethanol and kept in vacuum desiccator over P_2O_5 for 3 days. This was followed by a careful course of chemical analysis and melting point determination, to attain the exact preparation and the correct stoichiometry of the product was attained at.

b) *Ultraviolet Absorption Spectra Measurements*

In this investigation, the Nujol mull method was applied in order to obtain the ultraviolet absorption spectra of the different organic ligands and their transition metal complexes in solid dispersion before and after each gamma radiation dose. Thus, the crystal forms of the investigated specimens cannot be destroyed as in case of using the solvent technique. Accordingly, the dependence of ultraviolet absorption spectra on the the crystal morphology including lattice imperfection, degree of crystallinity which may be the result of gamma irradiation damage could be evaluated and examined in details. For this method, the powdered materials were mixed in an agate mortar with Nujol oil (A.R.) for 20 min. From these mixtures, transparent windows were made using Watteman's filter paper. In each run, a reference window, prepared without the material (plank), was used. The spectra was soonly automatically recorded using a Unicam SP 8000 ultraviolet recording spectrophotometer in the region 200 to 450 mμ. The spectrophotometer was adjusted at a suitable conditions of slit program, recording scale, speed and amplification which were constant throughout.

c) *Gamma Radiation Exposure*

For irradiation studies, the investigated specimen was irradiated using a ^{60}Co-gamma cell (activity 250 Ci) for absorbed gamma doses of 10^5 up to 10^9 rd. The spectral measurements were undertaken at room temperature and soonly within 15 min after removing the sample from the radiation field.

The spectra were further repeated three times and in each case, good repeatability of the result was ensured.

Results and Discussion

Fig. 1 shows the room temperature ultraviolet absorption spectra of the prepared (dbapH) and its corresponding metal complex of $Ni(dbapH)_2Cl_2$. Tab. 1 comprises the characteristic electronic absorption bands and their assignment before gamma irradiation. Thus, the electronic spectrum of (dbapH) exhibits the bands at 352 mμ and 278 mμ which are assigned to the pyridine and phenyl rings. The band at 342 mμ is assigned to the —CO group.

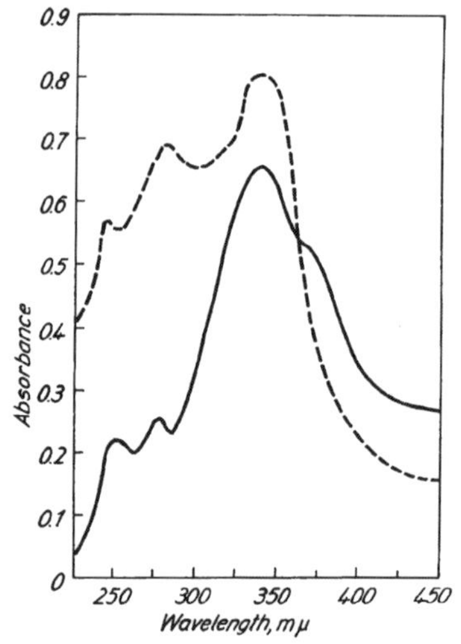

Fig. 1. The obtained ultraviolet absorption spectra of (———) dbapH organic ligand; (– – –) $Ni(dbapH)_2Cl_2$ metal complex before γ-irradiation

Tab. 1. The obtained ultraviolet absorption bands of (dbapH) ligand and its $Ni(dbapH)_2Cl_2$ solid complex before gamma irradiation

Compound	Py + Φ mμ	–C = 0 mμ
(dbapH)	252 and 278	342
$Ni(dbapH)_2Cl_2$	245 and 280	340

After complexation with Ni(II), there is no observed shift in the situation of the absorption bands whereas, there are noticeable changes in broadening and relative intensity of the bands.

Fig. 2 and Tab. 2 represent the obtained electronic (UV) absorption spectra of (daapH) ligand and its metal complexes $Co(daapH)_2Cl_2$ and $Ni(daapH)_2Cl_2$ before gamma irradiation. Thus, the electronic absorption spectra of (daapH) ligand display bands at 258 mμ due to pyridine

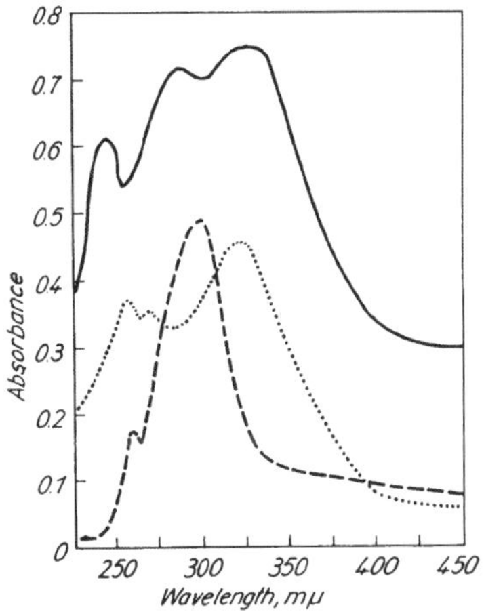

Fig. 2. The obtained ultraviolet absorption spectra of (– – –) daapH organic ligand; (· · · · ·) $Ni(daapH)_2Cl_2$; (———) $Co(daapH)_2Cl_2$ metal complexes before γ-irradiation

Tab. 2. Assignments of the obtained ultraviolet absorption bands of (daapH) ligand, $Co(daapH)_2Cl_2$ and $Ni(daapH)_2Cl_2$ solid complexes before gamma irradiation

Compound	Py mμ	–C = 0 mμ
(daapH)	258	302
$Co(daapH)_2Cl_2$	248 and 282	325
$Ni(daapH)_2Cl_2$	260 and 270	322

ring and at 302 mμ due to C=O group. After complexation, there is an appreciable shift toward higher wavelength amounting to 20 mμ for —C=O group. The band due to pyridine ring was found to split into two bands at 248 and 282 mμ and 260 and 270 mμ for $Co(daapH)_2Cl_2$ and $Ni(daapH)_2Cl_2$, respectively. Moreover, there are noticeable changes in broadening and relative intensity of the absorption bands upon complexation. These changes are ascribed to complex formation which takes place through creation of the chelate bonds between the transition metal ions (Co^{++} or Ni^{++}) and the donor centres of the involved organic ligands.

Fig. 3 and Tab. 3 show the effect of gamma irradiation doses on the electronic absorption spectrum of (dbapH).

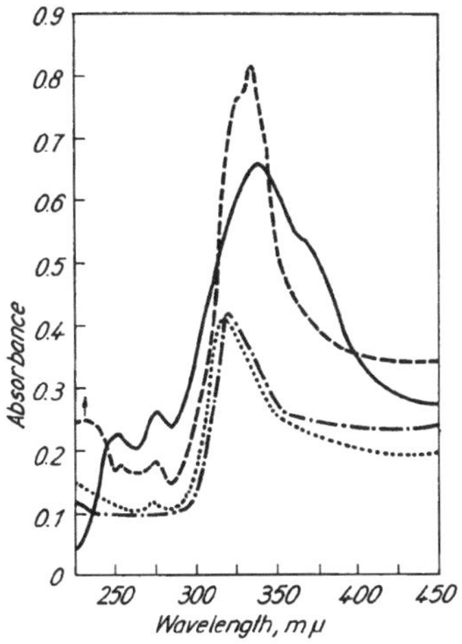

Fig. 3. The obtained room temperature ultraviolet absorption spectra of the prepared dbapH organic ligand as a function of γ-irradiation doses

(———) before γ-irradiation; (– – –) after the first dose (10^5 rd); (· · · · ·) after the second dose (10^7 rd); (· — · — ·) after the third dose (10^9 rd)

Tab. 3. Band assignments of the ultraviolet absorption spectra of (dbapH) ligand as a function of the absorbed gamma dose

Gamma radiation dose rd	Py + Φ mμ	–C = 0 mμ
0·0	252 and 278	342
10^5	231 and 275	333
10^7	– 274	322
10^9	–	316

It can be easily seen that as a result of gamma absorbed doses, the absorption bands suffer changes in their character, as occurred pronouncedly with energetic doses. These changes are summarized as follow:

i) The absorption bands at 250 and 280 mμ are gradually faden until disappeared completely at 10^9 rd.

ii) The absorption band at 342 mμ suffers blue shift up to 316 mμ as a function of energy of the absorbed gamma dose.

iii) Gamma induced absorption band is newly created at 225 mμ (see Fig. 3.).

Fig. 4 and Tab. 4 show the effect of gamma radiation doses on the ultraviolet absorption spectra of (daapH) organic ligand.

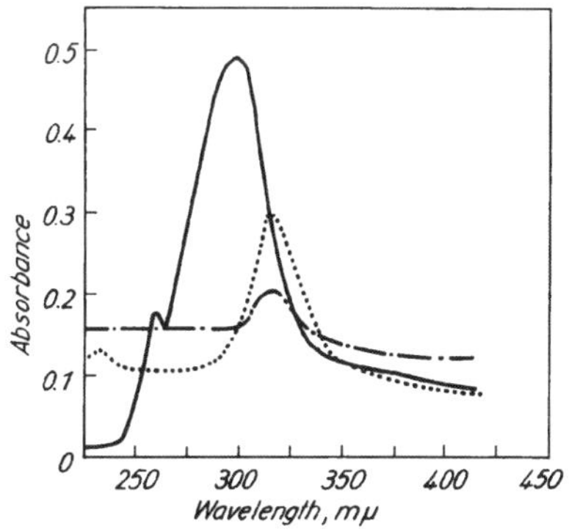

Fig. 4. The obtained room temperature ultraviolet absorption spectra of the prepared daapH organic ligand as a function of γ-irradiation doses

(———) before γ-irradiation; (· · · · ·) after the second dose (10^7 rd); (· — · — ·) after the third dose (10^9 rd)

Tab. 4. Band assignment of the obtained ultraviolet absorption spectra of (daapH) ligand at various gamma doses

Gamma absorbed dose rd	Py mμ	–C = 0 mμ
0·0	258	302
10^7	232	315
10^9	–	317

In this respect, the first radiation dose is not taken into account since its effect is still less pronounced than in the case of the prementioned (dbapH) ligand indicating its more resistance toward gamma radiation damage.

It is obviously seen that, the intensity of the absorption bands at 250 and 302 mμ are decreased gradually as a function of the absorbed dose energy indicating the possibility of such materials to be used as dosimeters.

Fig. 5 and Tab. 5 represent the effect of gamma radiation doses on the electronic absorption spectra of $Co(daapH)_2Cl_2$ solid complex. Thus, as a result of the gamma absorbed dose, the band at 280 mμ is disappeared completely by the first absorbed dose.

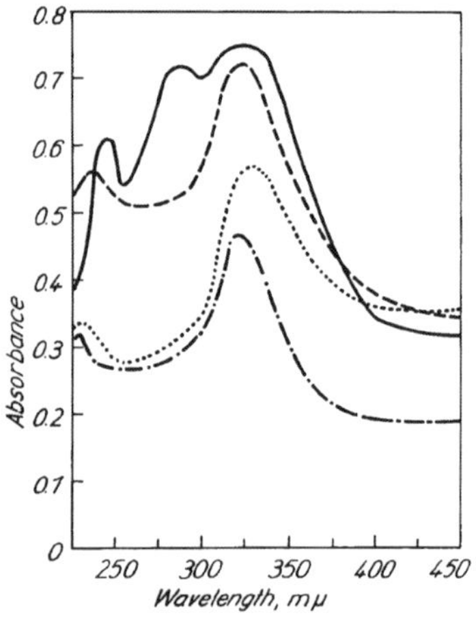

Fig. 5. The obtained room temperature ultraviolet absorption spectra of the prepared Co(daapH)$_2$Cl$_2$ metal complex as a function of γ-irradiation doses

(———) before γ-irradiation; (– – –) after the first dose (10^5 rd); (·····) after the second dose (10^7 rd); (·–·–·) after the third dose (10^9 rd)

Tab. 5. Band assignment of the obtained electronic absorption spectra of Co (daapH)$_2$ Cl$_2$ solid complex as a function of the absorbed gamma dose

Gamma radiation dose rd	Py mμ		−C = 0 mμ
0·0	248 and 282		325
10^5	235	—	320
10^7	230	—	332
10^9	227	—	320

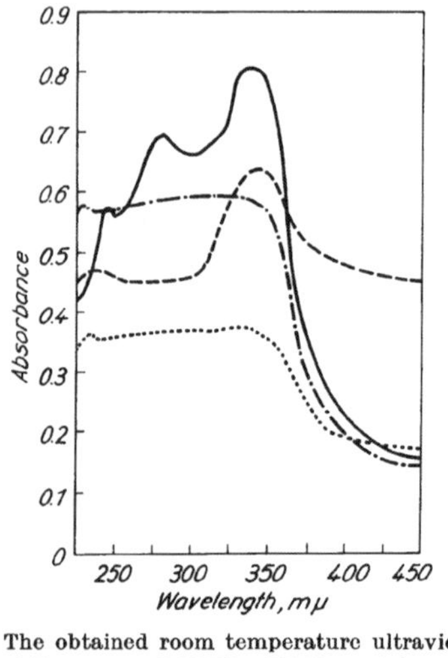

Fig. 6. The obtained room temperature ultraviolet absorption spectra of the prepared Ni(dbapH)$_2$Cl$_2$ metal complex as a function of γ-irradiation doses

(———) before γ-irradiation; (– – –) after the first dose (10^5 rd); (·····) after the second dose (10^7 rd); (·–·–·) after the third dose (10^9 rd)

Fig. 6 and Tab. 6 show the effect of gamma radiation doses on the electronic absorption spectra of Ni(dbapH)$_2$Cl$_2$ solid complex. Thus, as a result of gamma radiation, the band assigned to pyridine ring (280 mμ) disappeared completely at the first absorbed dose (10^5 rd). The band 340 mμ which is assigned to Co group disapperaed also at the second radiation dose (10^7 rd). In addition, there are noticeable changes in intensity and broadening of the other bands at higher extent than in (dbapH) ligand and its corresponding complex. This may be correlated with partial dissociation and free radical formation mechanism which is more liable to occur in (dbapH) ligand and its corresponding complexes. This indicates their lower stability and resistance towards gamma radiolysis as caused by the steric effect of substitution in (dbapH) ligand molecule.

Tab. 6. Band assignment of the electronic absorption spectra of Ni(dbapH)$_2$Cl$_2$ solid complex as a function of the absorbed gamma dose

Gamma absorbed dose rd	Py + Φ mμ		−C = 0 mμ
0·0	245 and 280		340
10^5	238	—	345
10^7	234	—	—
10^9	230	—	—

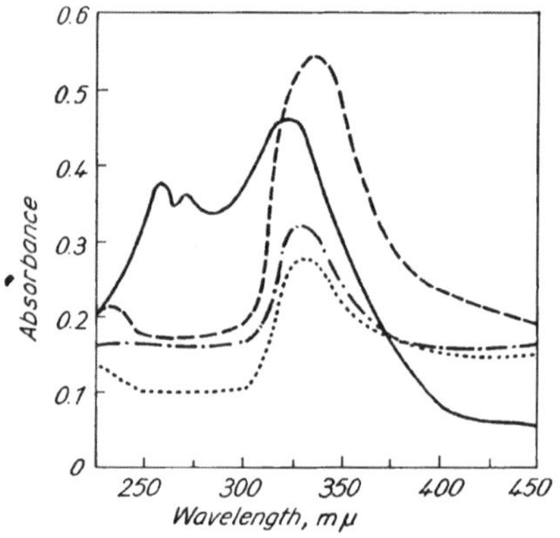

Fig. 7. The obtained room temperature ultraviolet absorption spectra of the prepared Ni(daapH)$_2$Cl$_2$ metal complex as a function of γ-irradiation doses

(———) before γ-irradiation; (– – –) after the first dose (10^5 rd); (·····) after the second dose (10^7 rd); (·–·–·) after the third dose (10^9rd)

Tab. 7. Band assignment of the electronic absorption spectra of Ni(daapH)$_2$Cl$_2$ solid complex as a function of the absorbed gamma doses

Gamma absorbed dose rd	Py mμ		−C = 0 mμ
0·0	260 and 270		322
10^5	253	—	337
10^7	226	—	332
10^9	—	—	330

Fig. 7 and Tab. 7 include the effect of gamma absorbed doses on the electronic absorption spectra of Ni(daapH)$_2$Cl$_2$ solid complex. Thus, the bands at 260 and 270 mμ due to pyridine ring disappeared at the third absorbed dose (10^7 rd). These notations illustrate that gamma radiation damage is more liable to occur on metal chelates than on their corresponding organic ligand molecules. This could be plausibly due to the weak chelate bond in relative to the regular chemical bonds (covalent bonds) in the organic ligand molecules.

General Discussion and Conclusion

In all cases as discussed herein, the decrease in the intensity of the electronic absorption bands as a function of gamma absorbed dose is correlated with the propagation of induced negative absorption in the same direction. As a

Kurznachrichten / Newlets

Conference report:
FOOD IRRADIATION GAINS MOMENTUM

With industrial interest at its highest level in decades, the technology of food irradiation processing is gaining momentum on the road to wide commercialisation. This conclusion can be drawn from reports presented at an international symposium on food irradiation processing held in early March in Washington DC. The meeting was sponsored by the IAEA and the FAO in co-operation with the US Government. Developments are far more positive than they've ever been, B.K. Wilson, of the radiochemical company of Atomic Energy of Canada Ltd (AECL) told the symposium. Although barries remain, he said, commercial interest in the process has been mounting in Canada, as in other countries.
The symposium drew considerable interest in the United States, where national food and health authorities have proposed regulatory changes that would allow wider use of the process, including irradiation of many fruits and vegetables. Addressing the symposium, Dr Sanford Miller of the US Food and Drug Administration, which regulates food safety, indicated that revised food irradiation rules are now nearing final approval and could be issued by this summer. Roughly 300 participants from some 40 contries attended the five-day symposium. An expert panel of international food irradiation authorities, convened at the symposium, made several recommendations aimed at fostering and accelerating the commercial development of the technology, as well as to focus future international co-operation in the field more sharply. The panel issued a final report containing following recommendations:
Programmes to achieve radiation disinfestation of stored foods should be expedited.
Programmes to achieve radiation disinfestation of fruit as a means of satisfying quarantine requirements should also be assigned a high degree of urgency.
The introduction of irradiation as a means of reducing food-borne disease resulting from contamination by pathogenic micro-organisms would also be important. The principal benefit would stem from the improvement in public health, although the promotion of international trade would also be of considerable importance. The establishment of regional aid programmes by international and national agencies should be given every encouragement. Potential users of food irradiation processing should be encouraged to bring to the attention of their national authorities the key role which could be played by the adoption of a uniform legislative framework for control of the process, based on the Codex General Standard. Every encouragement should be given to the production of educational and publicity material by national and international organisations and industry, with the aim of achieving the widest possible acceptance of the process by consumers. Additionally, the panel stressed that there is far more knowledge concerning the safety of food irradiation as a physical process, supported by many years of research and development, than is available concerning most other food preservation techniques. It is recommended that all government agencies be urged to recognise this fact, and to take all possible steps to facilitate its introduction on a commercial scale, so that this technology can be made available for the benefit of mankind. Regarding consumer acceptance, the panel agreed that public education and acceptance is an essential prerequisite for food irradiation development, and it urged the combined efforts of international agencies, governments, and industry to give the public accurate information by all available means. (Atom (1985) 343, S. 22-23, L. Wedekind)

VR Bulgarien:
ROENTGENFLUORESZENZANALYSATOREN IN DER VOLKSWIRTSCHAFT

In der VR Bulgarien wird unter Leitung des Instituts fuer NE-Metallurgie in Plovdiv intensiv an der Entwicklung und Einfuehrung von Systemen zur schnellen Bestimmung von Elementkonzentrationen in Rohstoffen, Halbzeugen und Fertigprodukten gearbeitet, die auf der Roentgenfluoreszenzmethode mit Isotopenanregung beruhen. Um den spezifischen Bedingungen bei den verschiedenen Anwendern zu entsprechen, werden drei Geraetetypen angeboten:
-Ein tragbarer 2-Kanal-Analysator mit Szintillationssonde und Kantenfiltern insbes. fuer den Feldeinsatz in der Geologie. Dieser Analysator ist z.B. fuer die Analyse von Pb und Zn sowie von Cu und Fe mit Pu-238 als Anregungsquelle ausgestattet.
-Ein 4-Kanal-Analysator, bevorzugt fuer den Einsatz in Erzaufbereitungen und Huetten zur operativen technologischen Ueberwachung. Als Zentralgeraet wird das 4-Kanal-Spektrometer NP 361 des Betriebes Gamma Muevek, Budapest, verwendet. Dieser Analysator ist zur Zeit fuer die Fe-, Pb- und Zn-Bestimmung in Blei- und Zinkkonzentraten und zur Analyse derselben Elemente, jedoch in anderen Konzentrationsbereichen, in Schlacken aus Schacht- und Fumingoefen der Bleichproduktion konzipiert.
-Ein Vielkanal-Analysensystem mit Halbleiterdetektor (gekuehlt), Vielkanalspektrometer und Computer sowie einem automatisierten Universalmesskopf mit den Strahlenquellen Fe-55 und Cd-109 oder Cm-244 und Am-241 fuer den Einsatz in betrieblichen Zentrallabors.
Mit diesem Analysator koennen alle Elemente des Periodensystems von Al bis U mit hoher Aufloesung bestimmt werden. Es liegt die Software fuer die Analyse zahlreicher Eisen- u. NE-Metallegierungen, von Staeuben, Granulaten, Schlacken, Stein und Pulpen vor. Der Einsatz dieser Geraete bringt einen hohen volkswirtschaftlichen Nutzen. Er liegt pro Geraet zwischen 52 000 und 202 000 Lw im Jahr.

Strahlentechnik:
STRAHLENCHEMISCH MODIFIZIERTE GUMMIELEMENTE

Ein Verfahren zur Oberflaechenmodifizierung fuer Gummidichtungen gestattet jetzt hoehere Belastungen beispielsweise durch Druecken, Temperaturen, Geschwindigkeiten und aggressive Medien, ohne dass Gummirezeptur und Herstellungstechnologie veraendert werden. Entwickelt wurde es vom VEB Cosid-Kautasit-Werke (Kombinat Plast- und Elastverarbeitung Berlin) und dem Institut fuer Technologie der Polymere der Akademie der Wissenschaften. Seine Besonderheit besteht darin, dass die Oberflaeche der Gummiteile mit beliebigem Profil eine duenne, fest verankerte Schicht erhaelt. Dank der chemischen Traegheit und geringen Adhaesion verleiht diese Schicht dem Ausgangsgummi eine Reihe wertvoller Gebrauchseigenschaften, so dass man von einer auf diese Weise geschaffenen neuen Materialklasse sprechen kann: den reibungs- und verschleissarmen Elastomeren. Mit diesem speziell entwickelten Verfahren der strahlenchemischen Modifizierung erreicht man: einen verringerten Haft- und Gleitreibungskoeffizienten in Abhaengigkeit von der Shore-A-Haerte des Gummis um 50 bis 70 Prozent des urspruenglichen Wertes, eine auf das sieben- bis 50fache erhoehte Verschleissfestigkeit, eine bedeutend erhoehte Alterungsbestaendigkeit, die voellige Beseitigung der stick-slip-Erscheinung bei geringen Translations- und Rotationsgeschwindigkeiten bis 0,003 m/s, besseres Beibehalten der Rueckstellkraefte und ein verbessertes Relaxationsverhalten. Derartig oberflaechenmodifizierte Dichtelemente haben sich z.B. in pneumatischen Schnellarbeitszylindern landwirtschaftlicher Maschinen und bei pneumatischen sowie hydraulischen Stellgliedern von Robotersystemen bewaehrt. Das Verfahren ist geeignet fuer die meisten in der Industrie eingesetzten und auf dem Weltmarkt gehandelten Gummidichtelemente.

(Spectrum 16(1985)7, S.V)

DURCH BESTRAHLUNG BESSERES BAUMATERIAL IN INDONESIEN

Als Ergebnis internationaler Zusammenarbeit konnte in der indonesischen Hauptstadt Jakarta eine Elektronenbestrahlungsanlage in Betrieb genommen werden, die der nukleartechnischen Entwicklung dieser Region und zugleich der dort stark expandierenden Baumaterialindustrie dienen soll. An der Verwirklichung des Projektes beteiligten sich die indonesische Regierung, die IAEA, das UNO-Entwicklungsprogramm (UNDP) und die Regierung von Japan. Die Anlage wird zur Bestrahlung von Sperrholz eingesetzt, wodurch sich die Qualitaet dieses Materials verbessern laesst. Aufgrund eines zwischen den Staaten im asiatischen und pazifischen Raum vereinbarten Abkommens (Regional Cooperative Agreement/RCA) kann die indonesische Elektronenbestrahlungsanlage auch durch die benachbarten Laender benutzt werden. Wie Generaldirektor Djali Ahimsa vom Indonesian National Atomic Research Board (Batan) anlaesslich der Inbetriebnahme der Anlage betonte, sind Radioisotope und Strahlentechniken in Indonesien bisher bereits auf anderen Gebieten eingesetzt worden, so zum Beispiel fuer geophysikalische und hydrologische Untersuchungen, zur Steigerung der Saatgutqualitaet und zur Haltbarmachung von Lebensmitteln. Der Aufschwung der Baumaterialindustrie erschliesst nun der Strahlennutzung eine weitere Anwendungsmoeglichkeit.[nach IAEA Bulletin, Spring, 1985] (SVA-Bulletin (1985)10)

Strahlenschutz:
NRPB-BERICHT UEBER DIE STRAHLENEXPOSITION DER BEVOELKERUNG

Das britische National Radiological Protection Board (NRPB) hat kuerzlich seinen 3. Bericht ueber die Bestrahlung der Bevoelkerung des Vereinigten Königreichs aus den verschiedenen natuerlichen und kuenstlichen Strahlenquellen veroeffentlicht. Entsprechende Studien hatte das NRPB bereits 1974 und 1978 herausgegeben. Der Ueberblick "The Radiation Exposure of the UK Population - 1984 Review" von J.S. Hughes and G.C. Roberts fasst die Situation in den fruehen 80er Jahren zusammen und kommt zu den folgenden Ergebnissen:
-Die mittlere jaehrliche effektive Aequivalentdosis aus allen Quellen zusammen betraegt fuer Einzelpersonen in der britischen Bevoelkerung rund 222 mrem (2,2 mSv).
-Der Hauptbeitrag zur Exposition der Bevoelkerung, naemlich rund 87%, geht auf natuerliche Quellen zurueck. Das NDPB hat die folgenden Anteile ermittelt: kosmische Strahlung 14%, terrestriche Gammastrahlung 19%, koerpereigene Strahlung 17%, Radon (Radon-222) 32%, Thoron (Radon-220) 5%.
-Aus kuenstlichen Quellen stammen 13% der Gesamtexposition. Der groesste kuenstliche Beitrag entsteht durch die Anwendung ionisierender Strahlen in der Medizin (11,5%). Verschiedene Quellen, einschl. Konsumgueter, Farbfernseher, Flugverkehr und die Verbrennung von Kohle, tragen 0,5% bei. Weitere 0,5% gehen auf das Konto des radioaktiven Fallout aus Atombombenversuchen in der Atmosphaere. Die berufliche Strahlenexposition tragt 0,4% zur mittleren effektiven Aequivalentdosis der britischen Bevoelkerung bei. Der kleinste Beitrag in der Erhebung des NRPB, naemlich 0,1%, stammt aus den Abgaben radioaktiver Stoffe aus Industrie- und Forschungsbetrieben, also einschl. Kernkraftwerken und Kernbrennstoffanlagen, an die Umwelt. (SVA-Bulletin (1985)10, S. 12-13)

Energiewirtschaft:
SONNENKRAFTWERK AUFGEGEBEN

Das staatliche franzoesische EVU Electricite de France (EdF) hat beschlossen, das vor zehn Jahren in Angriff genommene und 1983 ans Netz angeschlossene Sonnenkraftwerk "Thernis" bei Carcassonne in den Pyrenaeen aufzugeben. Die EdF begruendete diesen Beschluss mit den hohen Stromkosten des Sonnenkraftwerks. Die Anlage leitete die Sonnenwaerme mit 201 halbkreisfoermig angeordneten Spiegeln auf einen zentralen Kollektorturm. In den vergangenen Jahren waren zahlreiche Probleme aufgetreten. U.a. waren Spiegel von Windstoessen zerstoert und elektronische Steuerungen durch Blitzschlaege beschaedigt worden. (Atomwirtschaft/Atomtechnik 30(1985)7, S. K+U III)

Endlagerung:
ZERSTOERT STRAHLUNG DAS SALZ?

Das Phaenomen der Wechselwirkung von Kernstrahlung auf die Struktur und die Eigenschaften von Festkoerpern war seit Anbeginn der Kerntechnik ein ernstes Anliegen sowohl aus theoretischer als auch vor allem aus praktischer Sicht. Die in Form von Leerstellen von Kristallgittern (Frenkel- bzw. Schottky-Defekte) eingeleiteten Folgeprozesse wurden besonders intensiv an Natriumchlorid studiert, da sich hier die Entstehung von Farbzentren, Metallkolloidausscheidungen und Chlorfreisetzung mit verhaeltnismaessig einfachem analytischen Aufwand verfolgen laesst. Umfangreiche Untersuchungen seit Anfang der 50er Jahre an dieser Modellsustanz haben entscheidende Aufschluesse ueber das Verhaeltnis der strahleninduzierten Grundprozesse in Festkoerpern geliefert. Wenn jetzt in einigen neueren Arbeiten die Rede von bislang unbekannten oder vernachlaessigten Effekten ist, so stimmt dies mit der Wirklichkeit nicht ueberein. Die wichtigsten Parameter, die das strahlenchemische Geschehen beeinflussen, wie Temperatur, Strahlendosis, Dosisleistung und Fremdbestandteile im Steinsalz, sind schon lange bekannt. Eingeschlossen sind Erkenntnisse zum Bildungsmechanismus der Natriumkolloide durch Aggregation von F-Zentren. Dazu gehoert auch die experimentelle Beobachtung, dass sich bei hoeheren Temperaturen die Natriumkolloide wieder unter Rueckbildung von F-Zentren aufloesen koennen. In fast allen publizierten Arbeiten der letzten 30 Jahre kommt jedoch immer wieder zum Ausdruck, dass zusaetzliche und weiterfuehrende Untersuchungen zu noch nicht genuegend verstandenen Phaenomenen und Abhaengigkeiten vonnoeten sind. Dies gilt vor allem fuer sehr langsam ablaufende Prozesse und Sekundaereffekte, z.B. durch Rekombinationsreaktionen oder durch Verunreinigungen induzierte Abfangreaktionen. Die Sachverhalte werden dabei in der wissenschaftlich ueblichen vorurteilsfreien Art und Weise dargestellt. Insbesondere wird auf die durch experimentelle Defizite notwendigen Vorbehalten bei den teilweise gewagten Hochrechnungen hingewiesen. Mit den verfuegbaren Daten laesst sich aber eine Bewertung der moeglichen sicherheitsrelevanten Auswirkungen auf ein Endlager fuer hochradioaktive Abfaelle im Salzgestein vornehmen. Dramatisierende Darstellungen kommen dadurch zustande, dass mehrere Einzeleffekte mit ihren grossen Fehlerbandbreiten absichtlich so einseitig ausgewaehlt und miteinander verknuepft werden, dass man zu einer rund tausendfachen Uebertreibung des tatsaechlichen Geschehens gelangt. Zweifellos bestehen noch einige Detailfragen zur abschliessenden Quantifizierung des Strahleneinflusses. Zu ihrer Beantwortung bedarf es vor allem der schon lange geplanten In-situ-Versuche im Salzbergwerk ASSE mit echten HAW-Abfallgebinden. Fest steht allerdings schon heute, dass mit keinen unbeherrschbaren Sicherheitsproblemen zu rechnen ist. (Atomwirtschaft/Atomtechnik 30 (1985) 7, S. K+U III, E. Merz)

Kernenergie:
JUGOSLAWIEN RUND 6% ATOMSTROM

Das bisher einzige jugoslawische Kernkraftwerk Krsko hat im Jahre 1984 4,5 Mia kWh Elektrizitaet geliefert und damit zu rund 6% an die Gesamtelektrizitaetsproduktion des Landes beigetragen. Die Arbeitsausnutzung des KKW Krsko belief sich im vergangenem Jahr auf 76,5%, was im Vergleich mit den thermischen Kraftwerken Jugoslawiens einem der hoechsten Werte entspricht.(SVA-Bulletin (1985) 12, S. 8)

Reaktorsicherheit:

USA STUDIEN UEBER KERNSCHAEDEN VON TMI-2

Im Rahmen der laufenden Untersuchungen ueber den Zustand des beim technisch schwerwiegenden Stoerfall im amerikanischen Kernkraftwerk Three-Mile-Island-2 zerstoerten Reaktorkerns wurden kuerzlich in den amerikanischen Medien neue Hypothesen eroertert. Aufgrund vorbereitender Fernsehinspektionen des Reaktorinnern und anhand der Analyse von aus dem Reaktor entnommenen Brennstoffproben haben Forscher der EG&G Idaho im Auftrag des Department of Energy die Vorgaenge im Reaktorkern waehrend des Stoerfalls vom 28. 3. 1979 theoretisch rekonstruiert. Danach koennten die Brennstoffschaeden umfangreicher sein, als bisher angenommen wurde. Die EG&G Idaho postuliert, dass ein Teil des TMI-2-Brennstoffs im Verlauf des Stoerfalls zweimal schmolz und dass dabei hoehere Temperaturen aufgetreten sind als frueher berechnet. Mehr Klarheit wird sich im Lauf der kommenden Monate durch weitere Probeentnahmen und bei der Entfernung von Reaktoreinbauten und Brennstoff aus dem Druckgefaess ergeben. Im Zuge der Berichterstattung ueber die Arbeiten der EG&G Idaho wurde im April in einem Teil der Presse erneut faelschlicherweise behauptet, im Zusammenhang mit dem Stoerfall von 1979 sei in der Umgebung von TMI-2 erhoehte Kindersterblichkeit aufgetreten. Dies trifft nicht zu. Wie das zustaendige Gesundheitsministerium des Bundesstaates Pennsylvania, in dem das Kernkraftwerk Three Mile Island bei Harrisburg liegt, in umfassenden Untersuchungen festgestellt hat, kann kein Einfluss des Stoerfalls von TMI-2 auf die Kindersterblichkeit beobachtet werden. (+)

TMI-2: OBERES KERNGERUEST AUSGEBAUT

Als weiterer Schritt bei den Aufraeumungsarbeiten im amerikanischen Kernkraftwerk Three-Mile-Island-2 wurde Mitte 1985 das obere Kerngeruest aus dem Reaktordruckgefaess herausgehoben. Der Reaktordeckel war bereits im letzten Sommer vom Druckgefaess entfernt worden. Das 55t schwere obere Kerngeruest enthaelt u.a. die Steuerelementfuehrungsrohre und weist eine Hoehe sowie einen Durchmesser von je rund 3,5 m auf. Es ruht jetzt unter Wasser an seinem normalen, fuer Brennstoffwechsel-, Revisions- und Reparaturarbeiten vorgesehenen Abstellplatz. Nach dem Ausbau des Fuehrungsgeruestes ist jetzt der Reaktorkern freigelegt fuer die Entfernung des beschaedigten Brennstoffs aus dem Reaktor, mit der im Herbst 1985 begonnen werden soll. (SVA-Bulletin (+) (1985)9, S.9; (1985)12, S.12)

Veröffentlichungen / Publications

NEU ERSCHIENENE JAHRESBERICHTE

Jahresbericht 1984 des Instituts fuer Radiohydrometrie der Ges. fuer Strahlen- und Umweltforschung, Muenchen, April 1985, 217 S.

Scientific Report 1984 der Ges. fuer Schwerionenforschung, Darmstadt, March 1985, pp. 426

UNITED KINGDOM ATOMIC ENERGY AUTHORITY

AERE Bib 205
Publications list of the Environmental Safety Group Harwell laboratory 1971-1984 - Compiled J. Bromley; R.F. Emberton, January 1985, 61 pp.

AERE-M 3454
Evaluation of a sample introduction procedure to ICP-OES using graphite furnace atomisation. R.M. Brown; D.H. Nicholas, March 1985, 13 pp.

AERE-M 3461
An introduction to the ac impedance technique, and its application to corrosion problems - D.E. Williams; C.C. Naish, February 1985, 24pp.

AERE-R 11407
The retention of barium in man up to 19 years after intake, and its relevance to the metabolism of radium - D. Newton et al., January 1985 20pp.

AERE-R 11415
A literature review of EMATs, with stress calculations for simple EMAT geometries - J.A. Ogilvy, March 1985, 35 pp.

AERE-R 11475
Radioactive fallout in air and rain: results to the end of 1983 - R.S. Cambray et al., December 1984, 26 pp.

AERE-R 11481
Radiation damage to Si diode detectors by 14 MeV neutrons - D. West, November 1984, 8pp.

AERE-R 11496
A survey of the current and potenticl analytical techniques for the speciation of radionuclides in nuclear waste repository groundwaters and simulation leachates - P.M. Pollard, January 1985, 44 pp.

AERE-R 11564
A simple dislocation model for high temperature deformation - J.R. Matthews, January 1985, 37 pp.

AERE-R 11585
Modelling plutonium redistribution in fast reactor fuel pins - R. Thetford, December 1984, 23 pp.

AERE-R 11655
Thin films and interfaces. UKAEA conference on materials analysis by physical techniques. 32nd conference in the series formerly known as the Diffraction Analysis Conference - Edited B.A. Bellamy, A.E. Hughes, February 1985, 30 pp.

AERE-R 11660
The use of signal autocorrelation for the measurement of ultrasonic time-of-flight - D.R. Allen, April 1985, 27 pp.

Ausgewählte Beiträge / Recently published papers

ATOMNAJA ENERGIJA 59(1985)3 (in Russian)

Systems Analysis for Modal Control of Reactor Energy Distribution - O.L. Bozhenkov; P.T. Potapenko

On Radiation Hardening of Metals - I.V. Gorynin et al.

Methods of Studying Gas Evolution from Absorbents - I.G. Gverdtsiteli et al.

Radiolysis of Gadolinium Nitrate Aqueous Solutions - V.D. Ganzha et al.

The Choice of Material and Temperature Regime for the Target of a Fast-Ion Injector - V.G. Telkovsky et al.

Using the Concept of Radiation Capacity in Permissible Release Calculations - A.L. Kononovich

The Selection of Techniques for Estimating Pollutant in the Atmosphere when Standardizing Atmospheric Emissions of Radionuclides from Elevated Sources - N.E. Artyomova

Cs-137 Distribution in Samples of Fallout Data from a Number of Areas in the Baltic Sea During 1982-1983 - L.I. Gedeonov et al.

On the Effect of Corrosion of Fast-Reactor Fuel-Element Cladding on Its Mechanical Properties - E.F. Davydov et al.

Magnetic Resonance Studies of the Mobility of Lithium Ions and the Formation of Gamma-Radiolysis Products in Lithium Silicates - I.S. Pronin et al.

The Change in Volume Content of Deuterium in Heavy Water During Its Saturation with Carbon Dioxide - T.I. Efimova et al.

Microhardness of Neutron-Irradiated SiC(6H) Crystals - V.N. Brudny, B.T. Tolebayev

Data Bank for Reactor Material Test Methods - N.V. Markina et al.

ВНИИРТ - The Leading Science Center of Radiation Technique in the USSR - A.S. Shtan, E.R. Kartashev

IZOTOPTECHNIKA 28(1985)2 (all in Hungarian)
Experience with Computer Evaluation of RIA Measurements Based on the TECDOC-252 Publication of International Atomic Energy Agency - Frederic Joliot-Curie, L. Kocsar

Evaluation of RIA Investigations on Personal Computers - A. Booc

Studies on Modification of Ce-144 Tissue Deposition by Mixed-Ligand Complexes Treatment in Mice - A. Gachalyi et al.

The Diagnostic Value of Lymphoscintigraphy and Ultrasonography in the Examination of Abdominal Lymph Nodes of Patients with Lymphoma - L. Janoskuti et al.

The Natural γ-Background - Za. Nemeth, Sz. Berczi

The H-3 Labelling and Toxicokinetic Examination of Thiocarbamate Type Herbicids - L.Bako

Measurement of Cholecystokinin Content in Tissue and Plasma by Subtraction RIA - J. Nemeth et al.

A New and Neglected Possibility of Improving Our Health Physics Knowledge - L. Bozoky et al.

Health Physics in the Education - E. Viragh

IZOTOPTECHNIKA 28(1985)3 (all in Hungarian)

4th Congress of the Hungarian Society of Nuclear Medicine on isotope Diagnostics, 22-24 August, 1985, Sopron (Abstracts of papers)

J. LAB. CLIN. MED. 105(1985)1
Covalent binding of acetaldehyde to hepatic proteins during ethanol oxidation (using [C-14] ethanol) - V.A. Medina et al.

J. LAB. CLIN. MED. 105(1985)2
Hypochlorite radioiodination of parathyroid peptides - E. Orwoll et all.

Impairment of elastin resynthesis in the lungs of hamsters with experimental emphysema induced by sequential administration of elastase and trypsin (using [C-14]lysine) - M. Osman et al.

BRIT. J. NUTRITION 53(1985)1
Effects of meal consumption on whole body leucine and alanine kinetics in young adult men (using L-[1-C-13]$NaHCO_3$) - L.J. Hoffer et al.

Gluconeogenesis from caecal propionate in the horse (using [Z-C14]propionate and [2-^{3}H] glucose) - E.J.H. Ford, H.A. Simmons

Differences in intestinal protein synthesis and cellular proliferation in well-nourished rats consuming conventional laboratory diets (using L-[4-^{3}H]phenylalanine) - S. Southon et al.

Influences of cold exposure on digestion of organic matter, rates of passage of digesta in the gastrointestinal tract, and feeding and rumination behaviour in sheep given four forage diets in the chopped, or ground and pelleted form (using Ru-103-phenanthroline, [N-15]ammonium sulfate and Cr-51 labelled diet particles of 1mm) - P.M. Kennedy

Urea turnover and transfer to the digestive tract in the rabbit (using [C-14]urea and [N-15] urea) - S.J. Forsythe, D.S. Parker

ANGEW. CHEMIE 97(1985)5
Der Glycinrezeptor aus Rattenrueckenmark: Untersuchungen zum Wirkort des Pflanzenalkaloids Strychnin - H. Betz

Immobilisierung von radioaktivem Abfall im Festkoerper durch strukturelle Modifizierung der Matrix - Kristallchemie von Synroc, einer Titanat-Keramik - J. White et al.

Das 3'-Ende der tRNA und seine Rolle bei der Proteinbiosynthese - S. Chladek, M. Sprinzl

Photosignal II als Indikator fuer Waldschaeden - EPR-spektroskopische Untersuchung an Fichtennadeln (Nach Einwirkung von SO_2 und NO_x auf Fichtennadeln lassen sich EPR-Signalaenderungen in der Elektronentransportkette des Photosyntheseapparates nachweisen. Es kann zwischen Nadeln gesunder und geschaedigter Baeume klar unterschieden werden) - H.B. Stegmann et al.

J. ELECTRON SPECTROSC. RELAT. PHENOMENA 35(1985)1/2
Vibrational frequencies of isotopic species of $X-C{\equiv}C-H^+$ and $X-C{\equiv}C-X^+$, X= Cl, Br - D. Klapstein et al.

INORG. CHEM. 24(1985)3
Hydrogen isotope exchange between boranes and deuterated aromatic hydrocarbons : Evidence for reversible hydroboration of benzene - D.F. Gaines et al.

J. AMER. CHEM. SOC. 107(1985)5
Relative ^{36}S-^{34}S kinetic isotope effects - P. Paneth, W. Reimschuessel

^{34}S, ^{36}S and ^{18}O isotope effects on ^{31}P chemical shifts in thiophosphate anhydrides - C. Roeske et al.

J. VOLCANOLOGY AND GEOTHERMAL RES.23(1985)3/4
Geochemistry and isotopes of fluids from sulphur springs, Valles Caldera, New Mexico - F. Gott et al.

PRESSE DER SOWJETUNION 6(1985)5
Bereich der Zusammenarbeit - Dubna/Zeuthen - Hochenergiephysik - S. Kulinitsch

ENERGIETECHNIK 35(1985)4
Entwicklungstendenzen des brennstoff-energetischen Komplexes der UdSSR - S.M. Gortinski

WISSENSCHAFT IN DER UdSSR (1985)2
Kernwaffenkrieg, Atmosphaere und Klima - S. Baibakow et al.

Die Kernenergie veraendert die Chemie J.Kolotyrkin

ANGEW. CHEMIE 97(1985)4
Hochaufloesende Roentgen-Strukturanalyse-eine experimentelle Methode zur Beschreibung chemischer Bindungen (eine Uebersicht, 72 Lit.-Stellen) - K. Angermund et al.

Organo-Co-Verbindungen in der Pyridinsynthese - ein Beispiel fuer Struktur-Wirkungs-Beziehungen in der Homogenkatalyse - H. Boennemann

WISSENSCHAFT UND FORTSCHRITT 35(1985)5
Die Gefahr eines Nuklearkrieges (Alternativen-Konsequenzen) Teil II -

Max Volmer, Zum Gedenken an seinen 100. Geburtstag - K. Meyer

result of gamma irradiation, the photons may liberate some electrons. The free electrons and holes resulting from the irradiation process may be trapped at some of the intrinsic defects in the structure. The result is a decrease in the intensity of the electronic absorption. Another contributions to the negative absorption are most probably the dissociation, free radicals formation and polymerization mechanisms which may take place as a result of the interaction of gamma rays with the investigated materials giving rise to the formation of impurity centers and lattice defects.

As a result of gamma irradiation, there are a noticeable decrease in the main electronic absorption for the investigated Co(II) and Ni(II) complexes (particularly for $Ni(dbapH)_2Cl_2$ complex). But, this is not the case for their corresponding organic ligands. However, it may be possible to attribute this pronouncedly occurred negative electronic (UV) absorption to the reduction of either Co^{+++} or Ni^{+++} to their corresponding divalent cations (Co^{++}; Ni^{++}) in Co(II) and Ni(II) complexes, respectively as caused by gamma radiation interactions [11]. This was in conformity with the previous foundation of *Bishay* and *Kinawi* [12] for other materials.

Accordingly, results and discussion, on the electronic absorption spectra of materials investigated as a function of the absorbed gamma dose, provide us with a sequence of their arrangements in a descending order of stability against gamma radiation damage as follow:

$$\text{daapH} > \text{dbapH} > \text{Ni(dbapH)}_2\text{Cl}_2 > \text{Co(daapH)}_2\text{Cl}_2 > \text{Ni(daapH)}_2\text{Cl}_2$$

This indicated that gamma radiation damage highly affects on the relatively weak chelate bonds more than on covalent bonds within the organic ligand molecules. Steric hinderence due to substitution in the organid ligand may decrease the stability of the material and increase gamma radiation damage.

Received February 25, 1985

References

[1] *M. Frank, W. Stals,* Festkörper, Dosimetrie, Ionisierende Strahlung. Leipzig (1969), p. 198

[2] *A. E. Gillan, E. S. Stern,* An Introduction to Electronic Absorption Spectroscopy in Organic Chemistry. 2nd ed., Arnold, London 1957

[3] *J. W. Sidman,* Chem. Rev. **39** (1958) 689

[4] *K. Burger,* "Coordination Chemistry; Experimental Methods", Printed in Hungary 1975

[5] *H. Barbucci, G. Gialdi, G. Ponticelli, P. Paeletti,* J. Chem. Soc. (1969) 1775

[6] *H. Akaiwa, H. Kawamato, F. J. Izuni,* Inorg. Nucl. Chem. **37** (1975) 65

[7] *S. A. Bajue, C. C. Lalar, K. L. Stuart,* Ibid. **37** (1975) 51

[8] *R. Harrington,* Nucleonics **14** (1956) No. 9, 70

[9] *A. Bishay, A. Kinawi,* Proc. Intern. Conf. Physics Non. Cryst. Solids, North Holland Publ. Co., Amesterdam 1956

[10] *F. M. Ismail, E. M. H. Ibrahim,* Bull. N.R.C. **3** (1977) 27

[11] *A. Bishay,* Phys. & Chem. Glasses **2** (1961) 33

[12] *A. Bishay, A. Kinawi,* Proc. Intern. Conf. on Phys. Non-Cryst. Solids" North-Holland Publ. Co. Amsterdam 1965

Isotopenpraxis **21** (1985) 11, S. 391 bis 396

Entwicklung eines Extraktionsverfahrens zur Gewinnung von Technetium und Palladium aus der Abfallösung der Kernbrennstoffwiederaufarbeitung

Teil 3: Untersuchungen zur chemischen Stabilität und Strahlenresistenz von Tri-n-octylamin/Tetrachlorkohlenstoff

M. Gründig, H. Langguth, M. Richter

(Zentralinstitut für Isotopen- und Strahlenforschung der AdW der DDR)[1]

In Teil 1 und 2 [1, 2] wurde über Ergebnisse der Verteilungsuntersuchungen zur Extraktion von Tc und Pd mit Tri-n-octylamin (TOA) in CCl_4 berichtet. Angaben zur Selektivität des Trennprozesses wurden gemacht, ein Verfahrensschema wurde vorgeschlagen und der gesamte Prozeß in einer Laboranlage getestet. Im vorliegenden Teil wird die chemische Stabilität und Strahlenresistenz des Extraktionsmittels sowie der direkte Einfluß der Bestrahlung auf die Extraktion von Technetium und Palladium mit TOA in CCl_4 untersucht.

In part 1 and 2 [1, 2] we reported about the studies of distribution of technetium and palladium during the extraction with tri-n-octylamine (TOA) in CCl_4. Results about the selectivity were given, a flow scheme was proposed and the whole process was tested in a laboratory plant. In this part the chemical stability and the radiation resistance of the extraction agent and the direct influence of the radiation on the extraction of technetium and palladium with TOA in CCl_4 is investigated.

Keywords

fission products; palladium; radiation effects; radioactive waste reprocessing; solvent extraction; technetium

1. Einleitung

In Weiterführung der in Teil 1 und 2 [1, 2] beschriebenen Ergebnisse zur Extraktion von Palladium und Technetium mit Tri-n-octylamin wurde die chemische Stabilität und Strahlenresistenz des Extraktionsmittels untersucht.

Die Strahlenbeständigkeit und die chemische Stabilität eines Extraktionsmittels stellen wichtige Kriterien für die technische Anwendbarkeit bei der Kernbrennstoffwiederaufarbeitung dar. Bei längerer Nutzung des Extraktionsmittels im Zyklus sind die weitgehende Erhaltung seiner chemischen Zusammensetzung und damit seiner Extraktionsfähigkeit sowie die Beständigkeit des Lösungsmittels erforderlich. Um die Brauchbarkeit von TOA/CCl_4 für die

[1]) Anschrift: 7050 Leipzig, Permoserstr. 15

Tab. 1.

Radiolyseprodukte	Trennsäule	Säulenparameter			Dosiermenge	Trägergas	Empfindlichkeit	Detektor
		Länge m	Durchmesser mm	Temperatur °C	ml			
H_2	Molekularsieb 5A	2	4	Raumtemperatur	0,25	Argon	2-256	WLD
Methan, Äthan, Äthylen, n-Oktan	Carbowax 20 M auf Porolith	1,2	3	Raumtemperatur	0,50	N_2	$1:10^3$ $1:10^4$	FID einfach
Trioktylamin Dioktylamin Monooktylamin	Apiezon L auf Porolith mit KOH abgestumpft	2 Glassäule	3	150 – 300 °C 10 grd/min	10^{-4} $3 \cdot 10^{-4}$	N_2	$10^8 \cdot 30$	FID einfach

Wiedergewinnung von Spaltnukliden zu überprüfen, wurde die Einwirkung ionisierender Strahlung in Form von ^{60}Co-Gammastrahlung auf das Extraktionsmittel unter verschiedenen Bedingungen untersucht.

2. Experimentelles

Die Bestimmung der Verteilungskoeffizienten erfolgte, wie in [1] bzw. [2] beschrieben, unter Einsatz von ^{99}Tc/^{99m}Tc und ^{103}Pd. Die Palladiumnitratlösungen wurden wie in [1] erwähnt hergestellt, die Palladiumkonzentration wurde spektralphotometrisch mit Bismuthiol I bestimmt. Als Modellösung wurde die in [2] beschriebene verwendet.

Die Gamma-Bestrahlung der Proben erfolgte mit Dosisleistungen von 3 – 5 kGy/h im Dosisbereich von 10 kGy bis 1 MGy bei einer Temperatur von 25 °C und 40 °C. Es wurden 10-ml-Proben TOA bzw. TOA/CCl_4 als neutrale oder salpetersaure Lösungen in Spezialampullen, die mit gekapselten Gummistopfen verschlossen wurden, bestrahlt. In einer weiteren Versuchsserie erfolgte die Extraktion direkt während der Bestrahlung. Die Dosisleistung betrug hier bei Extraktion aus Salpetersäure 1,16 kGy/h und bei Extraktion aus Modellösung 0,85 kGy/h. Die Analyse der Radiolyseprodukte erfolgte gaschromatographisch, die Parameter sind in Tab. 1 zusammengestellt.

3. Ergebnisse

Da der Einfluß der Bestrahlung auf die Extraktion stark von den Bedingungen abhängt, unter denen das Extraktionsmittel bestrahlt wurde, wurden die Untersuchungen zur Aufklärung des Einflusses der Salpetersäure, des Lösungsmittels sowie der Bestrahlungsdosis in folgende Schritte gegliedert:

- Untersuchung der chemischen Beständigkeit des Extraktionsmittels TOA und 0,5 M TOA/CCl_4 gegen 3 N Salpetersäure ohne Einfluß von Strahlung
- Gamma-Bestrahlung von TOA, das mit 3 N Salpetersäure gesättigt wurde
- Gamma-Bestrahlung von 0,5 M TOA/CCl_4, das mit 3 N Salpetersäure gesättigt wurde
- Gamma-Bestrahlung von 0,5 M TOA/CCl_4 ohne Sättigung mit HNO_3
- Durchführung von Extraktionsversuchen mit diesen bestrahlten bzw. mit Salpetersäure vorbehandelten Extraktionsmitteln sowie einer Extraktion mit 0,5 M TOA/CCl_4 unter direkter Bestrahlung
- Untersuchung der gasförmigen und flüssigen Radiolyseprodukte mittels Gaschromatographie

Das System TOA/CCl_4 wurde in [3 – 6] ebenfalls hinsichtlich seiner chemischen und Strahlenstabilität betrachtet. Aus [3] ist bekannt, daß Lösungen von TOA in CCl_4 bei längerer Aufbewahrung und bei Bestrahlung Salzsäure freisetzen.

Die Veränderung der Farbe des Extraktionsmittels bei Bestrahlung zeigt Tab. 2.

Tab. 2. Farbänderungen des Extraktionsmittels TOA/CCl_4 nach Gamma-Bestrahlung

Dosis kGy	TOA/CCl_4 (0,5 M) ohne HNO_3 bestrahlt	TOA mit HNO_3 bestrahlt, mit CCl_4 verdünnt (0,5 M)	0,5 M TOA/CCl_4 mit HNO_3 bestrahlt
10	orange	hellgelb	farblos
30	orange	hellgelb	farblos
50	orange	hellgelb	schwach gelb
100	rot	hellgelb	schwach gelb
200	rot	hellgelb	schwach grün
500	braun	dunkelgelb	schwach grün
1000	braun	orange	orange, trüb gelbliche Gelteilchen

3.1. *Untersuchungen zur chemischen Beständigkeit des Extraktionsmittels*

Der Einfluß der Salpetersäure auf TOA bzw. TOA/CCl_4 wurde in Abhängigkeit von der Reaktionszeit ohne Gamma-Strahlung untersucht. Es wurde festgestellt, daß die Salpetersäure mit dem TOA Oligomere bzw. Polymere bildet, die abgetrennt werden konnten. Bei einer Reaktionszeit von 5 h beträgt die Polymerenbildung für TOA ca. 18 Gew.-% und für TOA/Tetra ca. 4 Gew.-% (Tab. 3). Mit diesen vorgesättigten Extraktionsmitteln wurde die Extraktion von ^{99m}Tc und ^{103}Pd durchgeführt. Es zeigte sich, daß die Kinetik der Einstellung des Gleichgewichts der Polymerenbildung für die Extraktion keine Rolle spielt, da schon nach wenigen Minuten konstante Verteilungskoeffizienten erreicht werden.

Tab. 3. Polymerenbildung des TOA bzw. TOA/CCl_4 unter Einwirkung von 3 N HNO_3

Extraktionszeit min	Polymerenbildung [Gew.-%]	
	des TOA	des 0,5 M TOA/CCl_4
5	10,5	–
15	12,0	0,8
30	13,0	1,7
60	14,9	2,5
200	17,0	3,6
300	17,5	4,0

3.2. *Radiolyse von TOA/CCl_4*

Der qualitative und quantitative Nachweis der gasförmigen und flüssigen Radiolyseprodukte des TOA erfolgte mittels Gaschromatographie. Als Radiolysegase wurden Wasserstoff, Methan, Äthan/Äthylen und n-Oktan nachgewiesen (Tab. 4–7). Bei Bestrahlung von reinem TOA ist Wasserstoff die Hauptkomponente des Radiolysegases. Methan, Äthan/Äthylen und n-Oktan entstehen nur in sehr geringer Menge (Tab. 4, 5). Bei Sättigung mit 3 N HNO_3 vermindert sich die strahlenchemische Ausbeute (*G*-Wert) der gasförmigen Radiolyseprodukte (Tab. 5).

Durch Reaktion des TOA mit der HNO_3 wird eine Stabilisierung erreicht, die den strahlenchemischen Abbau des Amins vermindert.

Die Zugabe des Lösungsmittels, bestrahlt wurden 0,5 molare Lösungen von TOA in CCl_4, bewirkt, daß die Bildung der gasförmigen Radiolyseprodukte weiter unterdrückt wird (Tab. 6, 7). Der stabilisierende Einfluß der Salpetersäure ist auch hier erkennbar (Tab. 7).

Äthan und Äthylen bilden sich nach einem radikalischen Mechanismus, was auf einen C—C-Bindungsbruch im TOA als auch auf den Zusammenschluß von CH_2-Radikalen zurückzuführen ist, während n-Oktan durch Spaltung der C—N-Bindung gebildet wird. In Spuren wurden Methyl- und Diäthylamin analysiert. Als Hauptkomponente der flüssigen Radiolyseprodukte wurde Oktylamin identifiziert. Dioktylamin entsteht nur in sehr geringer Menge (Tab. 8–11). Mit steigender Dosis verringert sich die strahlenchemische Ausbeute an Oktylamin beträchtlich. Zusatz von Salpetersäure übt wiederum einen stabilisierenden Einfluß aus. In Gegenwart des Lösungsmittels CCl_4 wächst die Menge der flüssigen Radiolyseprodukte, besonders des Oktylamins beträchtlich an. Dieser Effekt tritt hauptsächlich bei Abwesenheit von HNO_3 auf (Tab. 10). Die Anwesenheit von Tetrachlorkohlenstoff unterstützt also den strahlenchemischen Abbau des TOA. Bei Verwendung eines anderen Lösungsmittels kann somit noch eine Verbesserung der Stabilität des TOA erwartet werden. Aus der Analyse der gasförmigen und flüssigen Radiolyseprodukte sind in begrenztem Maße Rückschlüsse auf den Mechanismus der Reaktion zu ziehen. Bei der Radiolyse von reinem Tetrachlorkohlenstoff entstehen über einen Radikalmechanismus Chlor und Hexachloräthan [7].

Tab. 4. Gamma-Bestrahlung von reinem TOA, ohne 3n HNO_3 (gasförmige Radiolyseprodukte)

Probe	Dosis kGy	H_2			Methan			Äthan/Äthylen			n-Oktan		
		ml/l	Mol-% × 10^{-2}	*G*-Wert	ml/l	Mol-% × 10^{-2}	*G*-Wert	ml/l	Mol-% × 10^{-2}	*G*-Wert	ml/l	Mol-% × 10^{-2}	*G*-Wert
1	10	7,5	3,24	0,31	0,56	0,25	0,024	0,055	0,025	0,002	0,523	0,233	0,023
2	50	39,68	17,71	0,34	4,84	2,16	0,042	0,210	0,094	0,002	0,452	0,202	0,004
3	100	80,44	35,91	0,35	6,31	2,82	0,027	0,230	0,103	0,001	0,263	0,117	0,001
4	200	128,32	57,29	0,26	10,06	4,49	0,022	0,394	0,176	0,001	0,362	0,126	0,001
5	300	159,85	71,37	0,20	15,18	6,78	0,022	0,394	0,176	0,001	0,336	0,150	0,001
6	500	208,70	93,17	0,18	17,75	7,92	0,015	0,605	0,270	0,001	0,263	0,117	0,0002
7	1000	414,49	185,04	0,18	23,56	10,52	0,010	0,210	0,094	0,0001	0,209	0,093	0,0001

Tab. 5. Gamma-Bestrahlung von TOA, mit 3n HNO_3 vorbehandelt (gasförmige Radiolyseprodukte)

Probe	Dosis kGy	H_2			Methan			Äthan/Äthylen			n-Oktan		
		ml/l	Mol-% × 10^{-2}	*G*-Wert	ml/l	Mol-% × 10^{-2}	*G*-Wert	ml/l	Mol-% × 10^{-2}	*G*-Wert	ml/l	Mol-% × 10^{-2}	*G*-Wert
1	10	3,44	1,5	0,145	0,36	0,16	0,015	0,020	0,009	0,0008	1,104	0,493	0,048
2	50	18,28	8,2	0,158	2,02	0,09	0,017	0,053	0,024	0,0004	0,953	0,425	0,008
3	100	41,78	18,6	0,180	3,91	1,75	0,017	0,079	0,035	0,0003	0,760	0,317	0,003
4	200	70,52	31,5	0,152	6,84	3,05	0,015	0,184	0,082	0,0003	0,789	0,352	0,002
5	300	95,66	42,7	0,138	9,78	4,37	0,014	0,289	0,129	0,0004	0,467	0,208	0,001
6	500	98,56	44,0	0,085	10,06	4,49	0,009	0,473	0,211	0,0004	0,447	0,200	0,0003
7	1000	145,66	65,0	0,063	13,61	6,08	0,006	0,782	0,349	0,0003	0,460	0,205	0,0001

Tab. 6. Gamma-Bestrahlung von TOA/CCl_4 (0,5 molar) ohne HNO_3 (gasförmige Radiolyseprodukte)

Probe	Dosis kGy	H_2			Methan			Äthan/Äthylen			n-Oktan		
		ml/l	Mol-% × 10^{-2}	*G*-Wert	ml/l	Mol-% × 10^{-2}	*G*-Wert	ml/l	Mol-% × 10^{-2}	*G*-Wert	ml/l	Mol-% × 10^{-2}	*G*-Wert
1	10	0,19	0,085	0,008	0,09	0,04	0,004	0,001	0,0005	–	–	–	–
2	50	1,02	0,455	0,009	0,23	0,10	0,002	0,104	0,048	0,001	–	–	–
3	100	2,29	1,033	0,012	0,45	0,20	0,002	0,193	0,086	0,001	–	–	–
4	200	3,61	1,611	0,009	1,09	2,49	0,002	0,320	0,143	0,002	–	–	–
5	300	5,62	2,511	0,009	1,34	0,60	0,002	0,403	0,180	0,003	–	–	–
6	500	8,83	3,940	0,008	2,48	1,11	0,002	0,478	0,213	0,004	–	–	–

Tab. 7. Gamma-Bestrahlung von TOA/CCl_4 (0,5 molar) mit 3n HNO_3 vorbehandelt (gasförmige Radiolyseprodukte)

Probe	Dosis kGy	H_2			Methan			Äthan/Äthylen			n-Oktan		
		ml/l	Mol-% × 10^{-2}	*G*-Wert	ml/l	Mol-% × 10^{-2}	*G*-Wert	ml/l	Mol-% × 10^{-2}	*G*-Wert	ml/l	Mol-% × 10^{-2}	*G*-Wert
1	10	0,06	0,03	0,003	0,07	0,03	0,003	–	–	–	–	–	–
2	50	0,28	0,12	0,002	0,45	0,20	0,004	–	–	–	–	–	–
3	100	0,72	0,32	0,003	1,11	0,50	0,005	–	–	–	–	–	–
4	200	0,92	0,41	0,002	1,84	0,82	0,004	–	–	–	–	–	–
5	300	1,28	0,57	0,002	2,30	1,02	0,002	–	–	–	–	–	–
6	500	2,98	1,30	0,002	2,50	1,12	0,002	–	–	–	–	–	–

Tab. 8. Gamma-Bestrahlung von TOA ohne HNO_3 (flüssige Radiolyseprodukte)

Probe	Dosis kGy	Oktylamin			Dioktylamin		
		ml/l	Mol-% × 10^{-3}	*G*-Wert	ml/l	Mol-% × 10^{-3}	*G*-Wert
1	10	–	–	–	–	–	–
2	50	–	–	–	0,8	0,33	0,6
3	100	4,0	3,13	30	1,1	0,67	0,6
4	200	8,1	6,27	30	3,2	1,34	0,6
5	300	12,1	9,40	30	4,8	2,01	0,6
6	500	20,2	15,70	30	5,6	2,35	0,4
7	1000	59,1	45,80	44	6,4	2,68	0,3

Tab. 9. Gamma-Bestrahlung von TOA mit 3n HNO_3 behandelt (flüssige Radiolyseprodukte)

Probe	Dosis kGy	Oktylamin			Dioktylamin		
		ml/l	Mol-% × 10^{-3}	*G*-Wert	ml/l	Mol-% × 10^{-3}	*G*-Wert
1	10	–	–	–	–	–	–
2	50	21,0	16,29	31,5	7,9	3,35	6,5
3	100	29,9	23,19	22,4	8,8	3,72	4,2
4	200	36,4	28,20	13,6	12,7	5,37	2,6
5	300	38,0	29,45	9,5	15,9	6,71	2,2
6	500	43,2	33,50	5,8	24,0	10,16	1,9
7	1000	49,3	38,23	3,7	31,8	13,42	1,8

Tab. 10. Gamma-Bestrahlung von TOA-/CCl_4 (0,5 molar) ohne HNO_3 (flüssige Radiolyseprodukte)

Probe	Dosis kGy	Oktylamin			Dioktylamin		
		ml/l	Mol-% × 10^{-3}	*G*-Wert	ml/l	Mol-% × 10^{-3}	*G*-Wert
1	10	16,2	15,2	147,1	–	–	–
2	50	91,2	70,7	109,1	6,2	2,62	5,1
3	100	90,3	70,0	67,7	9,4	3,96	3,8
4	200	93,0	72,1	34,8	10,0	4,23	2,0
5	300	87,8	68,1	21,9	10,7	4,53	1,5
6	500	107,1	83,0	17,5	12,0	5,06	1,0
7	1000	168,3	130,5	17,6	15,8	6,68	0,6

Tab. 11. Gamma-Bestrahlung von TOA/CCl_4 (0,5 molar) mit 3 N HNO_3 behandelt (flüssige Radiolyseprodukte)

Probe	Dosis kGy	Oktylamin			Dioktylamin		
		ml/l	Mol-% × 10^{-3}	*G*-Wert	ml/l	Mol-% × 10^{-3}	*G*-Wert
1	10	–	–	–	–	–	–
2	50	9,9	7,65	14,8	–	–	–
3	100	11,8	9,15	8,8	–	–	–
4	200	19,1	14,80	4,9	4,3	1,81	0,9
5	300	23,6	18,33	3,9	9,4	3,96	1,3
6	500	34,1	26,40	3,7	18,2	7,70	1,2
7	1000	51,7	40,11	3,9	33,2	14,02	1,4

Weiterhin ist bekannt, daß bei der Radiolyse von HNO_3 zwei Mechanismen eine Rolle spielen [8].

$$HNO_3 \rightarrow HNO_3^* \begin{cases} \rightarrow \cdot OH + \cdot NO_2 & \text{Radikalprozeß} \\ \rightarrow 1/2\, O_2 + HNO_2 & \text{Molekularprozeß} \end{cases}$$

Bei Nanosekunden-Pulsradiolyse in reinem Tetrachlorkohlenstoff in flüssiger Phase wurden als Transienten das Radikalkation $CCl_4^{\cdot+}$, ein Ionenpaar-Neutralisationskomplex $CCl_4^{\cdot+} \ldots Cl^-$ und das Anion Cl_2^- beobachtet und identifiziert [9, 10]. Bei Zugabe von Alkanen, Alkenen, Aromaten bzw. Aminen (TOA) als Ladungsfänger findet ein Ladungstransfer des Radikalkations $CCl_4^{\cdot+}$ statt

$$CCl_4^{\cdot+} + TOA \rightarrow TOA^{\cdot+} + CCl_4$$

Das gebildete Radikalkation des TOA wird über eine mögliche Ionen-Molekül-Reaktion zu den stabilenEndprodukten Oktyl- und Dioktylamin weiterreagieren.

3.3. *Durchführung von Extraktionsversuchen mit den bestrahlten Extraktionsmitteln*

Abb. 1 zeigt die Extraktion von ^{99m}Tc und ^{103}Pd in Abhängigkeit von der Bestrahlungsdosis des mit 3 N Salpetersäure gesättigten TOA, das nach der Bestrahlung in CCl_4 gelöst wurde. Auf die Extraktion von ^{99m}Tc hat die Erhöhung der Bestrahlungsdosis keinen Einfluß. Der Verteilungskoeffizient für ^{103}Pd steigt bei Erhöhung der Dosis von 10 auf 1000 kGy auf das 10fache. Daraus ist ersichtlich, daß die entstehenden Radiolyseprodukte sich nicht negativ auf die Extraktion von Tc und Pd auswirken. Der Anstieg des Verteilungskoeffizienten für Palladium kann dadurch erklärt werden, daß bei der Radiolyse von Salpetersäure salpetrige Säure bzw. Nitritionen entstehen, die ähnlich wie Chloridionen die Extraktion von Palladium stark erhöhen.

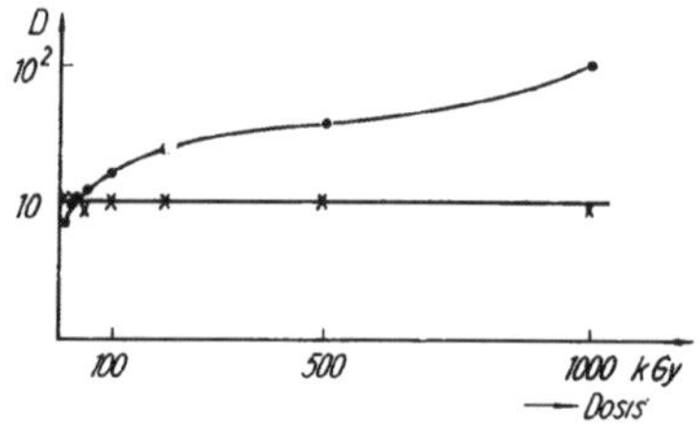

Abb. 1. Verteilungskoeffizient (*D*) von Technetium und Palladium mit bestrahltem Extraktionsmittel (TOA · HNO_3 bestrahlt) in Abhängigkeit von der Bestrahlungsdosis

c_{HNO_3} = 3,08 mol/l; c_{Pd} = 1,1 · 10^{-3} mol/l; —×— – D_{Tc}; —·— – D_{Pd}

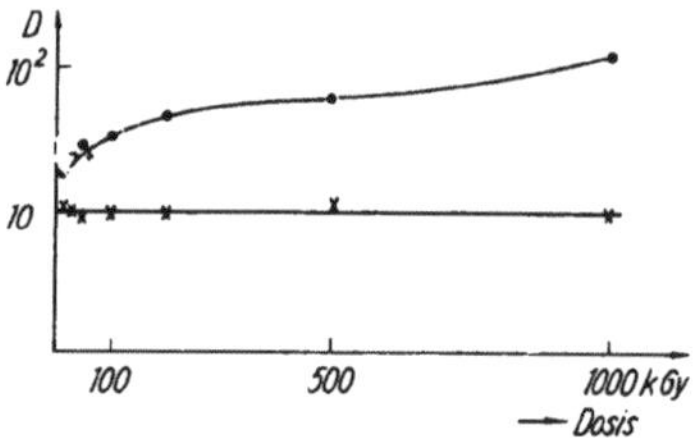

Abb. 2. Verteilungskoeffizient (*D*) von Technetium und Palladium mit bestrahltem Extraktionsmittel (TOA · HNO_3/CCl_4 bestrahlt) in Abhängigkeit von der Dosis

c_{HNO_3} = 3,08 mol/l; c_{Pd} = 1,1 · 10^{-3} mol/l; —×— – D_{Tc}; —·— – D_{Pd}

Abb. 2 zeigt die Extraktion von ^{99m}Tc und ^{103}Pd in Abhängigkeit von der Bestrahlungsdosis des mit 3 N Salpetersäure gesättigten TOA/CCl_4-Gemisches. Die Extraktion von ^{99m}Tc wird von der steigenden Dosis wiederum nicht beeinflußt. Die Verteilungskoeffizienten von ^{103}Pd liegen noch etwas höher als bei Abb. 1. Bei Bestrahlung von 0,5 molaren Lösungen TOA/CCl_4 vermindert sich die Bildung gasförmiger Radiolyseprodukte.

In [3] wurde die Konzentration der bei Bestrahlung frei werdenden Salzsäure im Dosisbereich 1–1000 kGy bei Verwendung 10%iger TOA-Lösungen in CCl_4 bestimmt. Der starke Anstieg der Verteilungskoeffizienten für Pd in Abb. 2 kann somit auf die wachsende Mengen Chloridionen bei zunehmender Dosis sowie auf vorhandene Nitritionen (analog Abb. 1) zurückgeführt werden.

In Abb. 3 und 4 ist die Extraktion von ^{99m}Tc und ^{103}Pd mit 0,5 M TOA/CCl_4 dargestellt. Das Extraktionsmittel wurde ohne vorherige Sättigung mit HNO_3 mit einer Dosis

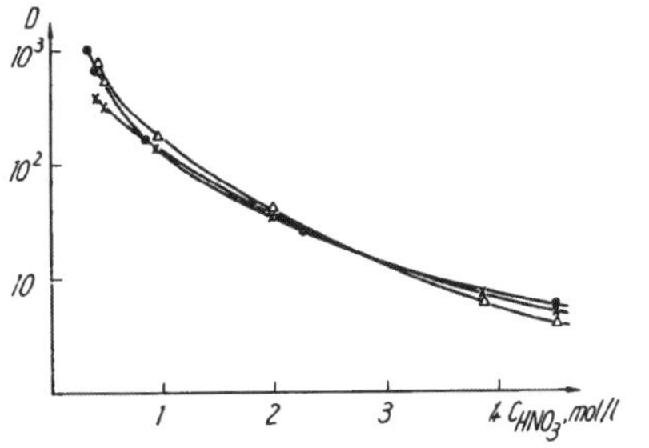

Abb. 3. Verteilungskoeffizient (D) von Technetium mit bestrahltem Extraktionsmittel (nicht mit HNO_3 vorgesättigt) in Abhängigkeit von der Salpetersäurekonzentration

c_{TOA} = 0,5 mol/l; —•— – unbestrahlt; —△— – Dosis = 10 kGy; —×— – Dosis = 1 MGy

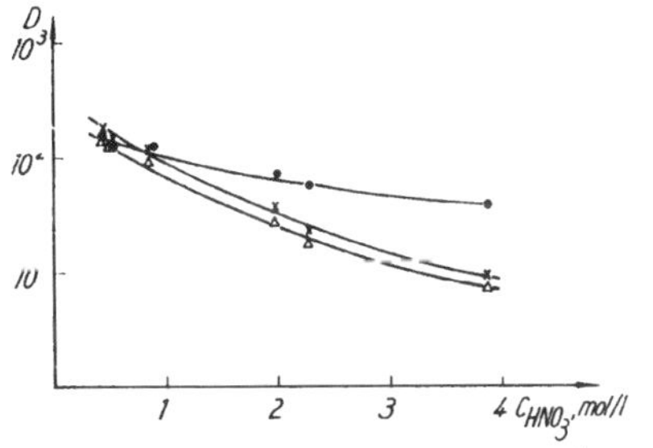

Abb. 4. Verteilungskoeffizient (D) von Palladium mit bestrahltem Extraktionsmittel (nicht mit HNO_3 vorgesättigt) in Abhängigkeit von der Salpetersäurekonzentration

c_{TOA} = 0,5 mol/l; c_{Pd} = 4,65 · 10^{-4} mol/l; —•— – unbestrahlt; —△— – Dosis = 10 kGy; —×— – Dosis = 1 MGy

von 10 und 1000 kGy bestrahlt. Abb. 3 zeigt, daß auch unter diesen Bedingungen die Verteilungskoeffizienten für ^{99m}Tc nahezu konstant bleiben. Bei der Extraktion von ^{103}Pd nehmen die Verteilungskoeffizienten bei höherer Säurestärke und Bestrahlung etwas ab. Das ist wie folgt begründet: Ohne vorherige Sättigung mit HNO_3 ist die strahlenchemische Ausbeute größer als bei Bestrahlung von TOA · HNO_3/CCl_4. Bei Anwesenheit von CCl_4 und Abwesenheit von HNO_3 erhöht sich z. B. der Anteil an gebildetem Oktylamin erheblich (Tab. 10). Weiterhin fehlt der entscheidende Einfluß der Nitritionen auf die Extraktion, da die Bestrahlung unter Abwesenheit von HNO_3 erfolgte.

3.4. *Durchführung von Extraktionsversuchen unter direkter Bestrahlung*

Die Extraktion unter direkter Bestrahlung kommt den realen Bedingungen der Spaltnuklidtrennung am nächsten. Deshalb wurde die Extraktion von ^{103}Pd und ^{99}Tc aus 3 N HNO_3 mit 0,5 M TOA/CCl_4 unter Bestrahlung durchgeführt, wobei eine Gesamtdosis von 330 kGy erreicht wurde. Aus Abb. 5 ist ersichtlich, daß bei beiden Systemen eine leichte Zunahme der Verteilungskoeffizienten erfolgt. Dieser Effekt kann z. T. auf kurzlebige Zwischenprodukte zurückgeführt werden, die in den bestrahlten Extraktionsmitteln bereits zu stabilen Endprodukten weiter reagiert haben. Für die Zunahme der Verteilungskoeffizienten für Pd können wiederum frei werdende Chlorid- und Nitritionen verantwortlich sein.

Abb. 6 zeigt die Extraktion von ^{103}Pd und ^{99}Tc aus einer Modellösung der beschriebenen Zusammensetzung [2]. Die erreichte Gesamtdosis betrug 219 kGy. Die Verteilungskoeffizienten zeigen für beide Nuklide einen ähnlichen Verlauf wie bei Extraktion aus Salpetersäure.

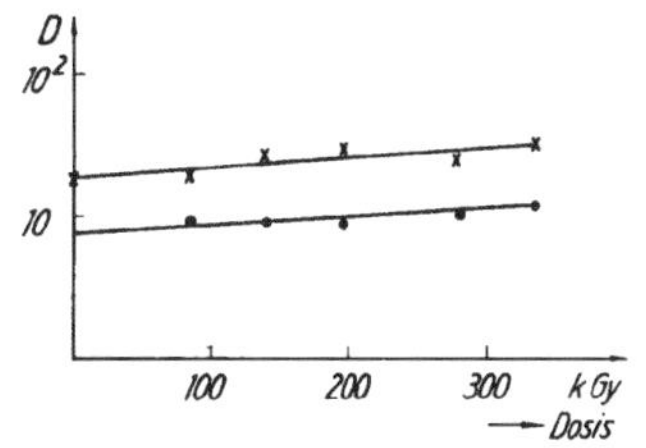

Abb. 5. Extraktion von Technetium und Palladium mit TOA/CCl_4 aus HNO_3 unter direkter Bestrahlung
Verteilungskoeffizient (D) von Technetium und Palladium in Abhängigkeit von der Dosis

c_{TOA} = 0,5 mol/l; P_D = 1,16 kGy/h); —×— – D_{Tc} (c_{Tc} = 1,93 · 10^{-4} mol/l; c_{HNO_3} = 2,92 mol/l; —•— – D_{Pd} (c_{Pd} = 3,56 · 10^{-3} mol/l; c_{HNO_3} = 3,05 mol/l)

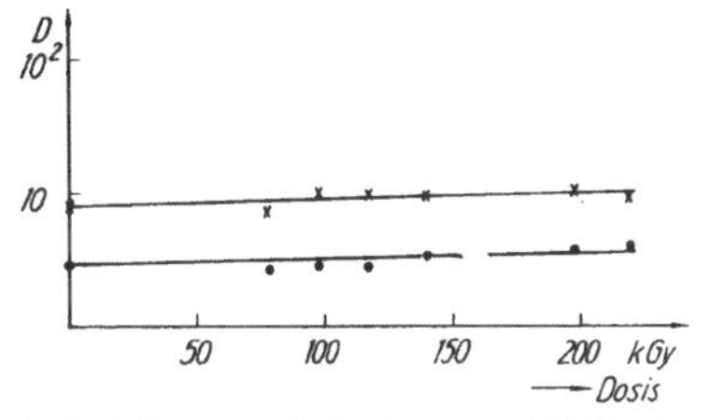

Abb. 6. Extraktion von Technetium und Palladium mit TOA/CCl_4 aus Modellösung unter direkter Bestrahlung
Verteilungskoeffizient (D) von Technetium und Palladium in Abhängigkeit von der Dosis

c_{TOA} = 0,5 mol/l; Dosisleistung = 0,85 kGy/h;—×— – D_{Tc} (c_{Tc} = 3,78 · 10^{-2} mol/l; c_{HNO_3} = 3,11 mol/l); —•— – D_{Pd} (c_{Pd} = 3,31 · 10^{-2} mol/l; c_{HNO_3} = 3,09 mol/l)

Es ist bekannt [11], daß der Einfluß der Bestrahlung von den Bedingungen abhängt, unter denen das Extraktionsmittel bestrahlt wird. Als günstig erweist sich in der Regel die Anwesenheit einer wäßrigen Phase, da in diesem Fall unerwünschte Radiolyseerscheinungen verschwinden, wahrscheinlich infolge der Entfernung störender Radiolyseprodukte durch die wäßrige Phase.

4. *Schlußfolgerungen*

Um die Ergebnisse von Strahlenresistenzuntersuchungen an den zur Trennung der Spaltprodukte benötigten Stoffsystemen in Bestrahlungsanlagen auf die realen Trennprozesse übertragen zu können, sind Abschätzungen der Dosisleistung im Apparat notwendig. Aus der Zusammensetzung der Abfallösung als Ausgangsprodukt für die Spaltproduktgewinnung und den resultierenden Energiedichten ist die Berechnung der Dosis, die das Trennmittel erhält, möglich.

Wenn wie beim vorliegenden Verfahren der Extraktion von Tc und Pd alle Komponenten, die wesentlich zur Aktivität der Abfallösung beitragen, in der wäßrigen Phase verbleiben, so daß bei der Reextraktion keine nennenswerte Strahlung auf die organische Phase einwirkt, beträgt die Gesamtdosis im Zyklus auf die organische Phase in einem definierten Extraktionsapparat für einen Durchsatz von 10 l/h 437 Gy [12]. Bei der von uns durchgeführten Extraktion unter direkter Bestrahlung zeigten die Verteilungskoeffizienten eine leichte Zunahme bis zu Dosen von 200 – 300 kGy. Das bedeutet, daß sich ca. 700 Zyklen ohne Zwischenregenerierung des Extraktionsmittels fahren lassen, wobei insgesamt etwa 6000 l Abfallösung verarbeitet werden können. Die Extraktionsversuche mit bestrahltem Extraktionsmittel zeigen, daß die entstehenden Radiolyseprodukte die Extraktion von Tc und Pd bis zu einer Gesamtdosis von 1 MGy nicht negativ beeinflussen. Danach könnte das Extraktionsmittel in ca. 80 Tagen 2300 Zyklen durchlaufen, wobei 19000 l Abfallösung verarbeitet werden können. Eine weitere Verbesserung der Stabilität des Extraktionsmittels kann offensichtlich erreicht werden, wenn Tetrachlorkohlenstoff durch ein anderes Lösungsmittel ersetzt wird. Eine Behinderung der Extraktion durch auftretende Feststoffe bzw. durch die Bildung einer 3. Phase unter Bestrahlung wurde nicht beobachtet. Das System Tri-n-octylamin in CCl_4 bzw. einem anderen Lösungsmittel erweist sich in dem untersuchten Dosisbereich als eine relativ strahlenresistente Substanz, so daß es für den Einsatz bei der Spaltproduktgewinnung geeignet erscheint.

Eingegangen am 2. 11. 1984

Literatur

[1] *B. Gorski, M. Gründig, M. Richter*, Isotopenpraxis **20** (1984) 93
[2] *M. Richter, B. Gorski, M. Gründig*, Isotopenpraxis **20** (1984) 97
[3] *F. Macasek, P. Rajec, V. Mikulay, R. Kopunec, R. Cech*, Materialy IV. Simpozyuma SEV, Issledovanija v Oblasti Pererabotki Obluchennowo Topliva, Vol. 2, Prague 1977, p. 188
[4] *F. Macasek, P. Rajec, V. Mikulay, R. Kopunec, R. Cech*, Coll. Czech. Chem. Commun. **43** (1978) 200
[5] *V. Mikulay, F. Macasek, P. Rajec*, Journal of Radioanalytical Chemistry **51** (1979) 55
[6] *F. Macasek, V. Mikulay, R. Kopunec, M. Korenovska, P. Rajec, P. Tarapcik*, Proc. ISEC-80/139, Vol. 3, 1 – 10 Liege, 1980
[7] *A. N. Zimin, S. S. Jegorowa*, in: *N. A. Bach*, Arbeiten über Strahlenchemie, Berlin: Akademieverlag 1960
[8] *K. Kaindl, E. H. Graul*, Strahlenchemie, Heidelberg: Dr. A. Huthig-Verlag GmbH 1967
[9] *R. Mehnert, O. Brede, W. Naumann*, Ber. Bunsenges. Phys. Chem. **83** (1979) 992
[10] *O. Brede, J. Bös, R. Mehnert*, Ber. Bunsenges. Phys. Chem. **84** (1980) 63
[11] *F. Baroncelli*, Nucl. Sci. and Engng. **17** (1963) 298
[12] *M. Richter*, Isotopenpraxis **17** (1981) 269

Isotopenpraxis 21 (1985) 11, S. 396 bis 399

Assessment of the Suitability of the New Track Detectors MA-ND/α and MA-ND/p for Investigation of the Distribution of Boron, Lithium, and Nitrogen in Steels

H. Firganek

(Institute for Metallurgy of Iron)[1]

In the paper the results of trials of implementation of new types of track detectors MA-Ni/α and MA-Ni/p are described for induced autoradiography of microadditions of B, Li and N in steels on a basis of the nuclear reactions:

$^{10}B(n, \alpha)^{7}Li, \quad ^{6}Li(n, \alpha)^{3}H, \quad ^{14}N(n, p)\,^{14}C$

The specimens have been examined in a casted state or after special technological treatments chosen in order to obtain the change in form and distribution of the analysed elements. Examination of B, Li and N distribution has been made with implementation of MA-ND/α, MA-ND/p and Kodak's CN-films CA 80-15 and LR-115. The special usability of track detectors MA-ND/p for investigations of distribution of N in steels containing ≦0,3% N has been found out. The autoradiograms of B and Li made on SSNTD MA-ND/α have similar quality as the autoradiograms made on the CN track detectors.

Es werden Versuchsergebnisse zur Nutzung der neuen Kernspurdetektoren MA-ND/α und MA-ND/p für die Untersuchung der Verteilung von Bor, Lithium und Stickstoff in Stählen angegeben. Es wurden Proben im Gußzustand, nach der plastischen Verarbeitung und der Wärmebehandlung untersucht. Die Wärmebehandlungsbedingungen wurden der Erkennung von Modifikationsänderungen der untersuchten Elemente angepaßt. Bei der Untersuchung der Verteilung von Bor, Lithium und Stickstoff wurden die Detektoren MA-ND/α, MA-ND/p und Cellulosenitrate der Firma Kodak Typ CA 80-15 und LR-115 eingesetzt. Es wurde eine besondere Eignung des Detektortyps MA-ND/p zur Untersuchung der Stähle mit niedrigem Stickstoffgehalt (≦0,3%) festgestellt.

Keywords

boron; impurities, lithium; nitrogen; photographic film detectors; spatial distribution; steels; trace amounts

Introduction

In the paper the results of trials of implementation of new types of track detectors MA-ND/α and MA-ND/p are described for induced autoradiography of microadditions of boron, lithium and nitrogen in steels on a basis of the nuclear reactions:

$^{10}B(n, \alpha)^{6}Li, \quad ^{7}Li(n\ \alpha)^{3}H, \quad ^{14}N(n, p)^{14}C$

In the work specimens of detectors have been used obtained from the Hungarian specialistes on XIII ISA in Tabor 1983,

[1]) Address: Institut Metalurgii Żelaza, ul. Miarki 12, 44-100 Gliwice/Polska

Experimental Conditions

The investigations have been made on specimens of steels and ferroalloys containing boron, lithium, boron and lithium together and nitrogen. The specimens have been examined in a casted state or after special technological treatments (plastic working, various types of heat treatment) chosen in order to obtain the change in form and distribution of the analysed elements.

The investigated materials have been complemented by specimen made from powder Li_2CO_3 and luminous paints containing the natural α-radioactive elements. The full set of investigated materials is shown in Tab. 1.

Tab. 1. Experimental materials

Material	Concentration of the invest. element		
	B	Li	N
1. Steel 25 HM	0,0001 – 0,003 % $10^{17} - 10^{18}$ at/cm³	–	–
2. Carbon-steel 15	0,001 – 0,003 % ~ 10^{18} at/cm³	–	–
3. Carbon-steel 15	–	0,0015 – 0,003 % ~ 10^{18} at/cm³	–
4. Carbon-steel 15	0,015 – 0,003 % ~ 10^{18} at/cm³	0,0015 – 0,003 % ~ 10^{18} at/cm³N	–
5. Li_2CO_3-powder	–	17 %	–
6. Nitriding Fe-Cr	–	–	5,4 % ~ 10^{22} at/cm³
7. Nitriding FeMn	–	–	3,8 % ~ 10^{22} at/cm³
8. Steel OH17AN-14M3	–	–	0,35 %B ~ 10^{20} at/cm³
9. Analtytical Standard of Steel	–	–	0,0062 % ~ 6 × 10^{18} at/cm³
10. Analytical Standard of Steel	–	–	0,0429 % ~ 10^{19} at/cm³
11. Analytical Standard of Steel	–	–	0,086 % ~ 10^{19} at/cm³
12. Analytical Standard of Steel	–	–	0,25 % ~ 10^{20} at/cm³

Examination of boron and lithium distribution has been made with implementation of MA-ND/α detectors and Kodak's CN-films of the type CA 80-15.

In the examination of nitrogen distribution the MA-ND/p detectors as well as have been implemented CA 80-15 and LR 115 Kodak's films.

It was assumed that due to the correctness or results of examination of boron distribution obtained on the CN-films veryfied during several years, the autoradiograms of boron made on these films may be used as reference points for assessment the new typs detectors.

The earlier results of examination of distribution of lithium and nitrogen have been not accessible, thus all types of detectors used for these elements have been implemented for the first time.

The constant parameter of experiments have been the irradiation conditions. All specimens have been irradiated during 5 minutes in the channel of the thermal column of the nuclear reactor EWA with following parameters:

the flux of thermal neutrons 0,025 eV – max. 4,3 × 10^{12} n/cm²s
the flux epithermal neutrons 0,49 eV – 6,3 × 10^{10} n/cm²s
$R_{Cd} - 5{,}7$
The total dose of thermal neutrons in all cases was 3 × 10^{14} n/cm².

This channel has been also used in earlier investigations of boron distribution due to the high flux of thermal neutrons; comparatively big loading diameter and possibility of realisation of short irradiation.

In order to compare the results the specimens have been irradiated in various sets applying among others:

- irradiation in one package the specimens of one type with various detectors,
- irradiation in one package various specimens with the some type of detectors,
- repetitive (in long intervals of time) irradiation of the chosen set of specimens with inplementation of various detectors.

The etching of detectors after irradiation has been carried out in water solution of NaOH changing their concentrations, temperature and etching time as well as method of detector immersion/two-side or one only the irradiation ide etching.

Detailed conditions of etching are shown in Tab. 2. The autoradiograms of luminous paints made on detectors MA-ND/α and MA-ND/p (etched in equal conditions like autoradiograms of the specimens irradiated in nuclear reactor) has been used like reference point in assessment of the influence of the etching conditions.

Analysis of the autoradiograms has been made by means of optical microscope, using magnification of 20 – 600 ×.

Criterions of the quality assessment of autoradiograms have been: tracks density of background and tracks of analysed element, diameters of tracks, contrast and resolution of image. As a whole about 100 autoradiograms have been analysed for each of the three examined elements.

Tab. 2. Etching conditions

	Element	Track detector	Solution	Temp. °C	Time h	Optimal conditions
1.	Boron	Kodak CA 80-15	10 % NaOH	20	3 – 10	10 % NaOH 20 °C, 5 h
	Lithium	MOM MA-ND/α	20 % NaOH 30 % NaOH 30 % NaOH	70 20 20	3 – 7 4 – 48 20 – 24	20 – 30 % NaOH, 20 °C, 24 h
2.	Nitrogen	Kodak CA 80-15	10 % NaOH 30 % NaOH 10 % NaOH	20 20 50 – 60	24 24 1	10 % NaOH 20 °C, 24 h
		Kodak LR 115	30 % NaOH	20	20 – 24	30 % NaOH, 20° C, 20 – 24 h
		MOM MA/ND/p	20 % NaOH 20 % NaOH 20 % NaOH	70 60 20	7 – 10 3 – 5 24 – 48 h	20 % NaOH 20 °C, 24 – 48 h
3.	Natural α-radioactive elements	MOM MA-ND/α MA-ND/p	20 % NaOH 20 % NaOH	70 20	1, 3, 5	20 % NaOH, 70 °C, 5 h

Experimental Results

The characteristic examples of autoradiograms of various types of specimens are shown on Fig. 1 – 11.

Analysis of the whole experimental materials led to the following conclusions:

- The density and diameter of tracks of natural α-radioactive elements on MA-ND/α and MA-ND/p detectors grows with growing of temperature of etching solutions and time of etching. The optimal quality of autoradiograms has been obtained when two-side etching has been made in conditions close to those proposed my manufacturer, e.a. 6 M NaOH, 70 °C, 5 h. The tracks of background of autoradiograms have been not noticed.

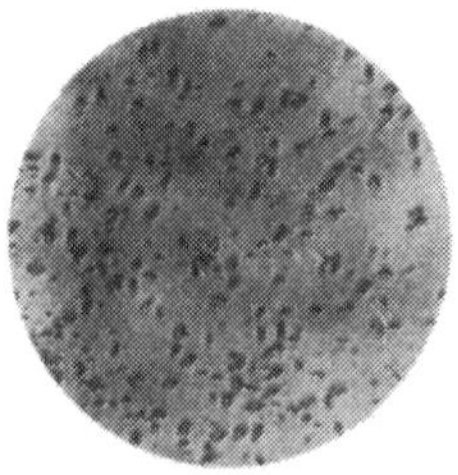

Fig. 1. Autoradiogram of the natural α-radioactive elements on MA-ND/ /α detector
Etch. 6 M NaOH, 70 °C, 5 h

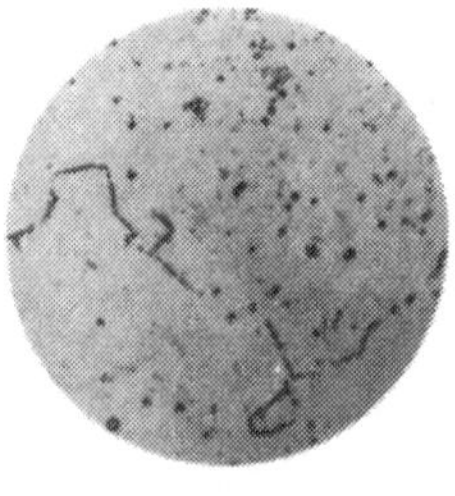

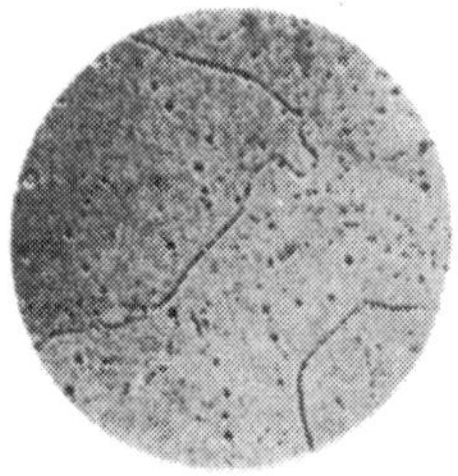

Fig. 2a Fig 2b

Fig. 2. Steel 15 (0,002 % B). Autoradiograms of Boron
a – Det. CA 80-15; b – Det. MA-ND/α

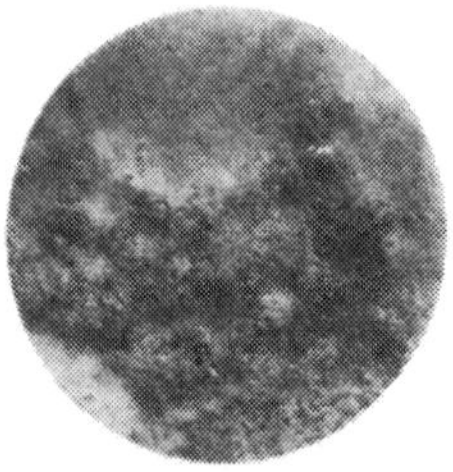

Fig. 3. Li_2CO_3 (powder). Autoradiogram of Lithium
Det. MA-ND/α

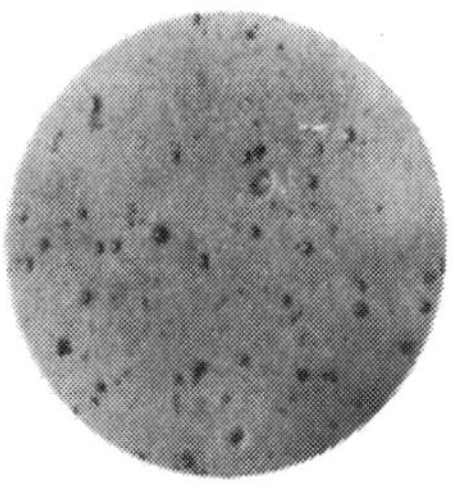

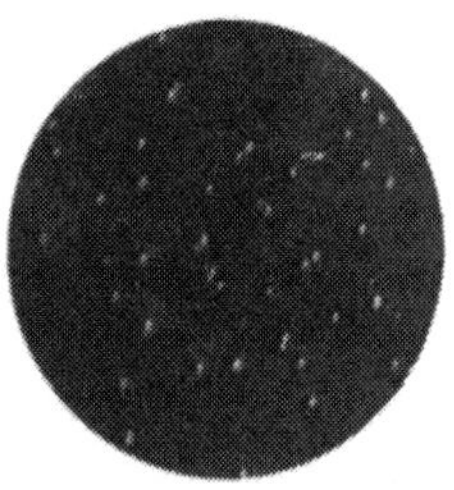

Fig. 4a Fig. 4b

Fig. 4. Steel 15 (0,003 % Li). Autoradiograms of Lithium
a – Det. CA 80-15; b – Det. MA-ND/α

Fig. 5 Fig. 6

Fig. 5. Steel 15 (0,003 % Li). Autoradiogram of Lithium
Det. MA-ND/α

Fig. 6. Steel 15 (0,003 % Li + 0,002 % B). Autoradiogram of Boron and Lithium
Det. MA-ND/α

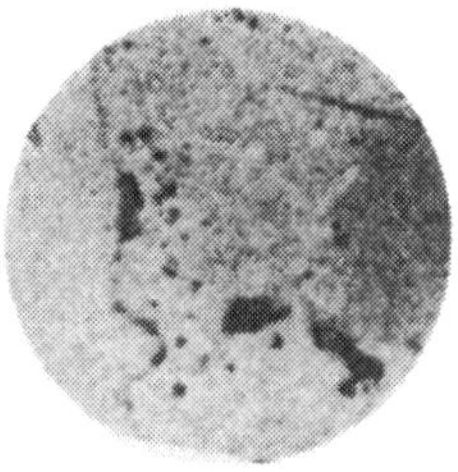

Fig. 7 Fig. 8

Fig. 7. Nitriding Fe-Mn (3,8 % N_2). Autoradiogram of Nitrogen
Det. MA-ND/p

Fig. 8. Steel OH17AN14M3 (0,35 % N_2). Autoradiogram of Nitrogen
Det. MA-ND/p

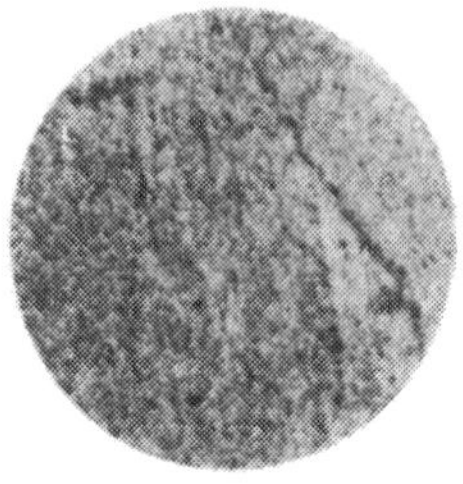

Fig. 9 Fig. 10

Fig. 9. Analytical standard of steel (0,0062 % N_2). Autoradiogram of Nitrogen
Det. MA-ND/p; Etch. 6 M NaOH, 70 °C, 5 h

Fig. 10. Analytical standard of steel (0,086 % N_2). Autoradiogram of Nitrogen
Det. MA-ND/p

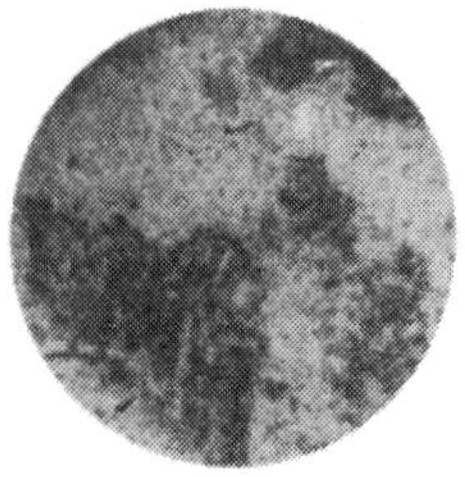

Fig. 11. Analytical standard of steel (0,25 % N_2) Autoradiogram of Nitrogen
Det. MA-ND/p

— The above noticed etching conditions has been found an improper for MA-ND/α and MA-ND/p detectors, if they are used for autoradiography of boron, lithium and nitrogen in specimens irradiated in nuclear reactor. The irradiation conditions created a big number of background tracks masking the real image of analysed element, especially when its concentration in examined material is very low, and distribution (caused by technological treatment) comparatively uniform. In the case of boron it applies to the specimens quenched in high temperature with big fraction of boron in solution.

In these conditions for all three examined elements much better contrast and legibility of autoradiographical image/ without real loss of sensitivity) have been obtained etching the detectors in 20—30% NaOH, 20 °C, during 24 h, or even more (autoradiograms of nitrogen) only from oneside.

The changing of etching conditions results in distinct reduction of the density of background tracks having not in the some time the higher influence on reduction of quantity and diameter of tracks of the nuclear reaction

$$^{10}B(n, \alpha)^7Li\,,\quad ^6Li(n, \alpha)^3H\,,\quad ^{14}N(n, p)^{14}C\,,$$

products.

It is caused by physical and chemical properties of B, Li and N, their inclination to segregation in defepcted areas and compounds which they made with other components of steel.

They are then the elements the distribution in steel of which is naturaly heterogenous (even if they are in solution) and tracks recorded on detectors are more easy detected because they represent microareas with comparatively big local concentration of the components.

The diameter of B, Li and N tracks on MA-ND/α and MA-ND/p detectors measured from several 3—20 μm and density of tracks were 10^2—10^6/cm². It has been found that the differences in size and number of tracks have been normaly connected with technological history of material and not with the conditions of irradiation and etching of detectors.

The autoradiograms of B and Li made in the conditions of irradiation and etching of MA-ND/α detectors used in this work are comparatively to autoradiograms of these elements made on CN-films, being slightly more sensitive for Li. It concerns as well the average density of background tracks as the size of the set of specimens made from the some steel to which the microadditions of B, Li and both of them have been added (with an assumption, that the two elements would made the some sompounds) shown that the differences in size and number of tracks resulting form addition the second element was similar for the two types of detectors; in each case making it possible concluding from which steel it comes out.

As fur as the autoradiography of nitrogen is concerned the results of experiments acknowledge the high sensitivity of MA-ND/p detectors in comparison with CN-films. These differences are less visible in investigation of specimens with higher (>1%) concentration of nitrogen and reveal them selves especially when specimens containing <0,3% of nitrogen are examined. Using MS-ND/p detector it has been possible to obtain ever the autoradiogram of analytical standards of steel containing 0,0062% N_2, in spite of the fact then in this case due to the low concentration and comparatively uniform distribution of nitrogen on the autoradiogram only the contour of specimen is visible slighty were intensive then the background.

To conclude the work it way be stated that the properties of the examined detectors guarantee their usability for investigation of boron, lithium and nitrogen distribution in steels containing microadditions of these elements in the conditions when the only possible neutron source is nuclear reactor.

Received January 7, 1985
Accepted in revised form April 3, 1985

Isotopenpraxis 21 (1985) 11, S. 399

Buchbesprechung

Gas Hydrates

E. Berecz, M. Balla-Achs. Budapest: Akadémiai Kiadó and Amsterdam: Elsevier Science Publishers B. V. 1983, 343 S., 225 Abb., 66 Tab., 587 Lit. (Studies in Inorganic Chemistry Vol. 4); karton. M 87.50

Die Autoren behandeln nach einem kurzen Überblick über die historische Entwicklung der Literatur über die Gashydrate zunächst die Grundlagen der Struktur der Hydrate und dann die Bedingungen für die Bildung und Zersetzung einschließlich des Einflusses von weiteren Gaskomponenten, von Elektrolytzusatz zum Wasser sowie die Rolle der Inhibitoren. In den nächsten Kapiteln werden die thermodynamischen Eigenschaften derartiger Systeme (formelmäßige Beziehungen, statistisch-thermodynamische Untersuchungen) sowohl in allgemeiner Form als auch in Form spezieller Daten konkreter Systeme dargestellt. Dies betrifft u. a. die Hydratsysteme mit Stickstoff, Sauerstoff, Edelgasen, Halogenen, Kohlendioxid, Schwefeldioxid, Kohlenwasserstoffen, Gasgemische aus diesen Komponenten und auch Naturgasen. Im letzten Teil des Buches stehen die technologischen Aspekte sowie die verschiedensten Methoden der experimentellen Untersuchung von Gashydrat-Systemen im Vordergrund.

Klar und übersichtlich sind die theoretischen, praktischen und technologischen Gesichtspunkte dieser wichtigen Klasse von Verbindungen in den einzelnen Abschnitten dargelegt. Die Struktur, die Eigenschaften und die Thermodynamik der Gashydrate werden verständlich beschrieben und durch viele Detailangaben konkretisiert. Aus der angebenen Anzahl von Diagrammen und Tabellen ergibt sich ein bemerkenswert großer Vorrat an Werten und sonstigen Angaben über diese Systeme. Das ebenfalls umfangreiche Literaturverzeichnis sowie der Autor- und Subjekt-Index ermöglichen eine leichte Handhabung dieses m. E. gelungenen Buches, das auch für Spezialisten angrenzender Wissensgebiete wertvoll sein dürfte. *P. Harting*

Isotopenpraxis 21 (1985) 11, S. 400 bis 404

On Problems of Standardized Handling and Quantitative Evaluation of Autoradiograms*)

H. C. Treutler, K. Freyer

(Central Institute of Isotope and Radiation Research, Leipzig)[1])

In the last years autoradiography has gained greater importance as a quantitative method of measuring radioactivity or element concentration. Usually a relative measurement is used. The optical density of the photographic emulsion produced by a calibrated irradiation source is compared with that produced by a sample. The influence of different parameters, such as β-particle energy, backscattering, fading of the latent image, developing conditions, matrix effects and others on the results and the errors of the quantitative evaluation of autoradiograms are described. This method is demonstrated in the case of the quantitative determination of the gold content in silicon.

In den letzten Jahren wurde die Autoradiographie zunehmend als quantitative Methode zur Radioaktivitätsmessung bzw. zur Bestimmung von Elementkonzentrationen genutzt. Meist wird sie als Relativmessung eingesetzt. Die Schwärzung der Fotoemulsion durch eine geeichte Strahlenquelle wird mit der von einer Probe verursachten Schwärzung verglichen. Es werden die Einflüsse verschiedener Parameter wie Energie der β-Teilchen, Rückstreuung, Fading des Latentbildes, Entwicklungsbedingungen und Matrixeffekte auf Ergebnis und Fehler der quantitativen Auswertung der Autoradiogramme beschrieben. Die Methode wird am Beispiel der quantitativen Bestimmung des Goldgehaltes in Silizium demonstriert.

Keywords

accuracy; autoradiography; calibration; calibration standards; errors; measuring methods; photographic film detectors; sensitivity

Introduction

Autoradiography is often regarded as a "qualitative" method of investigation, the desired information being whether there is more radioactive material present in certain structures than in others. However, this very simple requirement really involves a quantitative decision, because the word "more" implies a comparison in magnitude between two quantities. Autoradiography is concerned with subjective or objective comparisons of qualities. First of all the amounts of optical densities or of silver grain densities are compared [1].

Herrmann and coworkers [2] investigated the possibilities, limitations and errors of quantitative autoradiographic methods. They used very thin radioactive labelled polymer foils as samples for their investigations. The exposure and development conditions were standardized very carefully, i.e., the exposure and the development were thermostated to within ±0·1 °C, the developer and the fixing bath were very clean and fresh, the movement of the film in the developer was standardized. Under there ideal conditions the error of the quantitative autoradiography was determined. It is lower than ±10%.

But in most of the practical autoradiographic investigations the experimental conditions are not so perfect. It is necessary to compare autoradiograms exposed and developed under very different conditions. Often the exposure and development dates are far apart.

Another important question which must be answered is the following: when we compare the optical densities of different structures in the same or in different autoradiograms, do we really compare the total amounts of radioactivity within the structures or do we compare the particle flux density or the fluency which leaves the sample surface? In either case it is necessary to define the extent in space within the sample. If the distributions of the radioactive labelled substances within the different structures are of similar depth, the problem is of minor importance. But should we compare, for example, the amount of radioactivity of a substance deeply distributed in the sample with that in a thin surface layer corrections are necessary.

In a previous paper [3] it was demonstrated that the optimum of standardization of autoradiograms with the aim to quantify the autoradiographic investigation is the relative method. The aim of this paper is the elaboration of a method for the quantitative evaluation of autoradiograms and the calculation of the errors of this method under real conditions.

1. Standardization of autoradiograms

Many parameters influence the relation between the optical density in the developed emulsion (D) and the particle flux density emitted by a sample (Φ):

$$D = f(\Phi t) = c_s \cdot c_d \cdot \Phi \cdot t \tag{1}$$

with c_s — sensitivity factor of the emulsion

c_d — factor of development conditions (influence of temperature and age of developer, developing time, fixation, washing, drying)

t — exposure time

It is practically impossible to calculate exactly all these parameters, but it is necessary to know this relation, for

*) Presented at the XIVth International Symposium on Autoradiography, 19–23 November 1984, Reinhardsbrunn, GDR.

[1]) Anschrift: 7050 Leipzig, Permoserstraße 15

Tab. 1. Data of the autoradiography calibration set (ACS)

radionuclide	half life	maximum energy	particle flux density*)
	a	MeV	s^{-1} cm^{-2}
^{63}Ni	100	0·067	(2·5 ± 0·5) 10^6
^{99}Tc	2·1 · 10^5	0·292	(2·68 ± 0·18) 10^4
^{204}Tl	3·8	0·760	(1·4 ± 0·2) 10^4
^{90}Sr/^{90}Y	28·5	0·54/2·27	(2·16 ± 0·14) 10^4

*) The particle flux densities were determined by the Amt für Standardisierung, Meßwesen und Warenprüfung (ASMW) of the GDR.

example, in a series of autoradiograms to compare optical densities of one autoradiogram with those of another.

For this purpose it is useful to expose all autoradiograms with an autoradiography calibration set (ACS) as described in [4, 5]. This ACS consists of four foil samples radioactive labelled with ^{63}Ni, ^{99}Tc, ^{204}Tl, ^{90}Sr/^{90}Y (see Tab. 1) on a plastic holder. On the understanding that the ACS is exposed for the same time and the intervals between the dates of exposure are very small in comparison to the half life of the radionuclide the optical densities in the autoradiograms should be equal. But the differences in sensitivity of the photographic emulsion of different film sheets, the different development conditions and last not least the error of the densitometric measurement result in a deviation of the measured optical densities. These influences taken together affect also the measured optical densities of other samples exposed on this film sheet. To standardize a series of autoradiograms the deviations of the optical densities of the ACS will be used as correcting factors to calculate the real optical densities of samples in the following way:

1. Measurement of the optical densities D_i of the ACS in all of the n autoradiograms
2. Calculation of the mean value of these optical densities

$$\bar{D} = \frac{1}{n} \sum_{i=1}^{n} D_i$$

3. Determination of the correcting factor $f_i = \frac{\bar{D}}{D_i}$
4. Multiplication by f_i of all optical densities measured in the i-th autoradiogram of the samples under investigation.

For the estimation of the amount of these deviations the ACS was exposed ten times on ORWO-AF3 and six times on ORWO-TF14 autoradiographic film. Also the development of these films took place independently under real conditions without thermostated developing baths and at different dates. The optical densities of the samples of the ACS in the autoradiograms were measured five times in each case by means of a Zeiss-Schnellphotometer G II. The mean values of all autoradiograms for each of the four samples were normalized to 100%.

Fig. 1 shows the deviation of the optical densities ΔD of every sample from the mean value for independent exposures. This investigation demonstrates the variation of the relation between the optical density in the developed emulsion and the particle flux density emitted by a sample under real conditions. Normally this deviation amounts to about ±10%. But it is also possible (Fig. 1a, exposure 10) that the deviation is as much as ±30%.

The standardization of autoradiograms is the first step in quantitative autoradiography. In the way described above the influences of the developing conditions and partly of the sensititity changes in the photographic

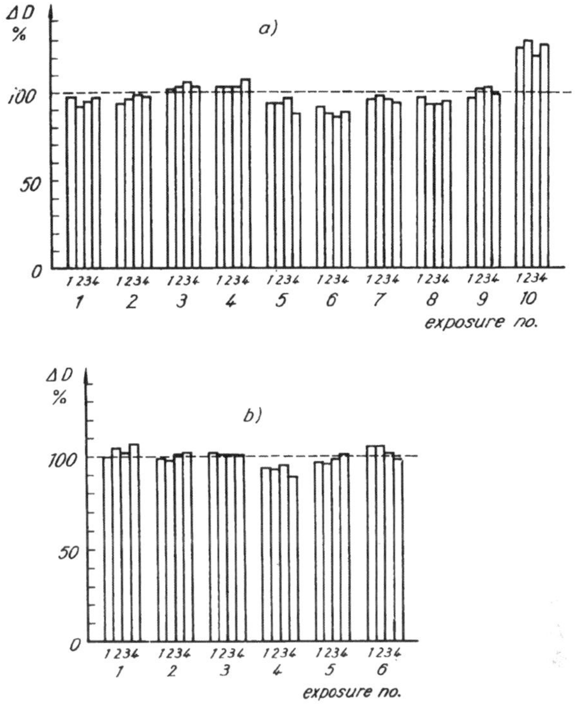

Fig. 1. Deviation of optical densities of the same samples from the mean value of ten (fig. 1a) and six (fig. 1b) independent exposures
a) autoradiographic film ORWO-AF 3; 1-^{63}Ni; 2 – ^{99}Tc; 3 – ^{204}Tl; 4 – ^{90}Sr/^{90}Y
b) autoradiographic film ORWO-TF 14; 1 – ^{63}Ni; 2 – ^{99}Tc; 3 – ^{204}Tl; 4 – ^{90}Sr/^{90}Y

emulsion are corrected. Thus also the optical densities measured in different types of emulsion will be comparable with each other.

2. Calibration of autoradiograms

The second step is the calibration of autoradiograms. In the simpler case that means estimating the relation between the standardized optical density and the particle flux density or fluency. But in most cases the amount of radioactivity in the different structures of the sample or the concentration of the radioactive labelled element in these structures is wanted.

In practice, the first case is the estimation of the value of the sensitivity factors in equation (1). The sensitivity splits into two parts. The first one is included in the standardization of autoradiograms (c_d) and contains the influence of the change of the sensitivity properties of the photographic emulsion depending on the conditions of production storage and development. The second part (c_s) depends on the quality and the energy of the ionizing particles. The sensitivity of an autoradiographic emulsion is the reciprocal of the number of particles per unit area producing the optical density $D = 1$. By means of the ACS the sensitivity of autoradiographic emulsions can be determined quantitatively for these four radionuclides because the particle flux densities are known (Tab. 1).

The answer to the question for the amount of radioactivity or the concentration of the radioactive labelled element is more complicated. Information about the depth distribution of this element in the sample and matrix properties must enter into the calibration because it is necessary to take into consideration the influences of scattering, backscattering, and absorption of the ionizing particles in the matrix. These influences can be considered with sufficient accuracy by the method of comparison only. The principles of this method are described in [6] by the example of phosphorus diffusion profiles in silicon. In this way the activation conditions are also included in the relative calibration. As a relative real standard a sample is used in which the element concentration is laterally homogeneous. This concentration is known with a high accuracy. The depth distribution of this element corresponds to that in the investigated sample [7]. This real standard is handled in the same way as the sample from activation to exposure. The standardized optical densities are calibrated in units of element concentration in this way.

Practical experience with phosphorus diffused silicon samples as real standards for quantitative autoradiographic investigations of semiconductor basic material is described in [7]. The mean error of these measurements was $\pm 25\%$. Without the errors of the concentration measurement of the standard.

3. Example of a quantitative autoradiographic investigation

The problem was to determine the diffusion profile of gold in silicon by means of layer-by-layer chemical etching. The lateral distribution of gold was very inhomogeneous in these samples. Therefore the radioactivity measurement in the etching solution or in the etched silicon samples gave incorrect results which gave information on a "mean profile" only. With the autoradiographic method it was possible to determine the depth distribution in connection with the local coordinates. The procedure of this investigation is shown schematically in Fig. 2.

Neutron activation in a nuclear reactor

The gold diffused samples were activated together with a gold standard (AuCl) of known quantity and with phosphorus diffused silicon standards described in [5] in the nuclear research reactor Rossendorf.

Chemical preparation. During the time the radioactivity of the silicon (^{31}Si) is dying away a part of the activated gold standard was deposited on the clean surface of a pure silicon sample. By measuring the radioactivity of the gold standard relative to that of the deposited gold on the silicon sample the number of gold atoms per unit area was determined. This gold sample was used as gold coated silicon standard and was calibrated in units of gold atoms per cm².

Thickness measurement. The thickness of the gold diffused samples was determined mechanically using a TESA-MODUL equipment.

Autoradiography. All autoradiographic exposures were done together with the ACS to standardize the optical densities of all exposures. The first autoradiogram shows the original lateral surface distribution of gold in silicon.

Chemical etching

Using a suitable etchant solution a thin layer of the gold diffused sample was removed.

Thickness measurement. The thickness of the removed layer was determined as the difference between the sample thickness before and after chemical etching.

Autoradiography. The autoradiogram shows the distribution of the residual gold in silicon together with the listed standards.

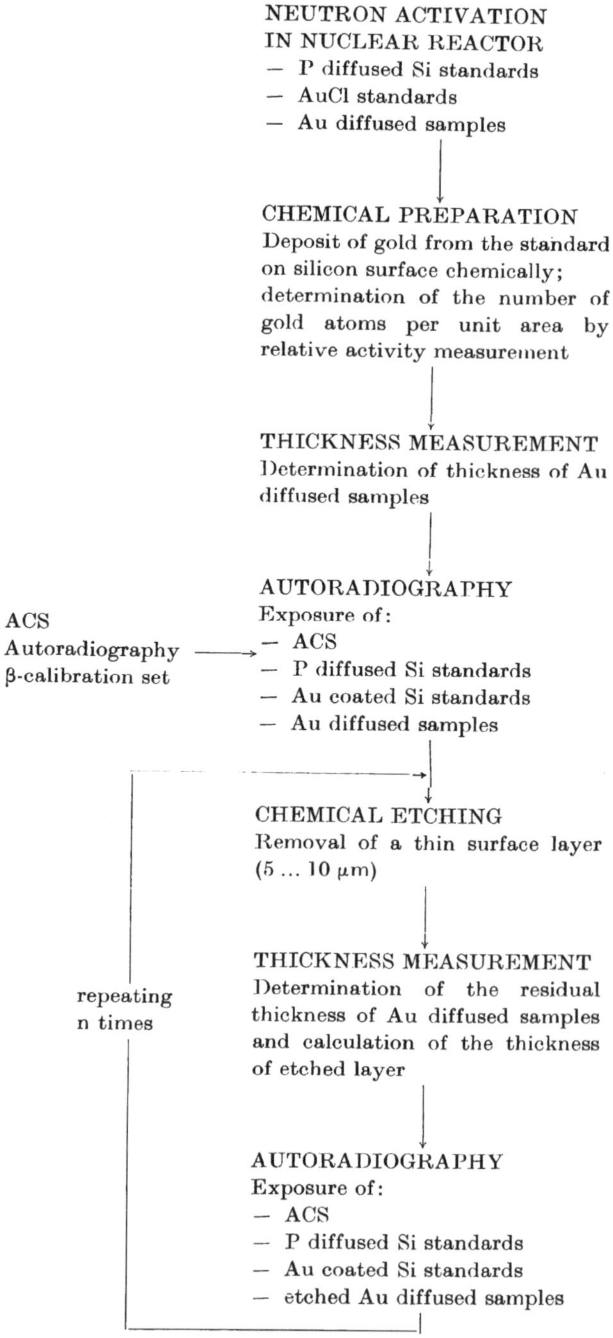

Fig. 2. Scheme of the procedure of investigation to determine the depth distribution of gold diffused in silicon

This handling (chemical etching – thickness measurement – autoradiography) was repeated to a depth of about 100 μm. The residual radioactivity of the gold became lower in the period of investigation of more than one week as a result of the decay with a half life time of 2·7 d and of the lower concentration of gold at greater depths. Consequently the last autoradiograms were produced on the highly sensitive autoradiographic film ORWO-TF14, the first autoradiograms on ORWO-AF3. Fig. 3 shows one of these standardized and calibrated autoradiograms.

By the help of the standard autoradiograms the optical densities of the gold diffused samples can be transformed into element concentration values.

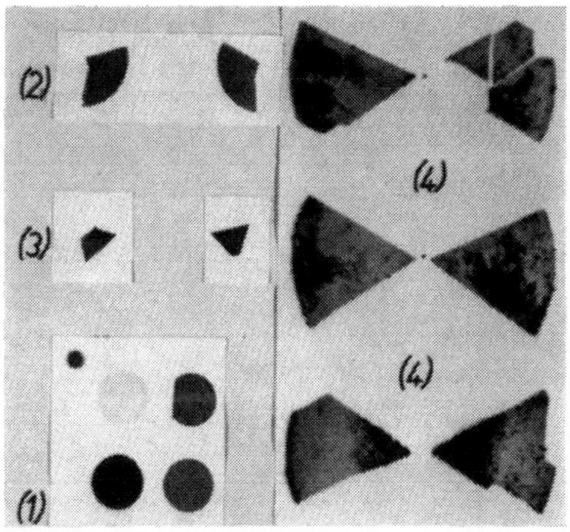

Fig. 3. Copy of one autoradiogram of the described investigation
(1) – ACS; (2) – phosphorus diffused standards; (3) – gold coated standards; (4) – gold diffused samples

To estimate the measurement errors of this method the measured optical densities of the gold coated standards and of the phorphorus diffused standards were standardized using the correcting factor f_i determined by means of the ACS and corrected with reference to the half lives of these radionuclides. The relative standard deviations of the single measurement ΔD and those of the mean values ΔD_m of ten (ORWO-AF 3) and six (ORWO-TF 14) single measurements are collected in Tab. 2 dependent on the film type. The relatively high errors of the optical densities of gold coated samples are attributed to the optical densities of two of these exposures with values $D > 1{\cdot}2$. In this region the density curve (particle fluency per unit area vs. optical density) is not linear (see Fig. 4). So it would be necessary to correct the density values.

Tab. 2. Relative standard deviations of the standardized optical densities
ΔD – relative standard deviation of the single measurement
ΔD_m – relative standard deviation of the mean values

film type	ΔD		ΔD_m	
	phosphorus	gold	phosphorus	gold
ORWO-AF 3	±4 %	±18 %	±1·3 %	±6·3 %
ORWO-TF 14	±4 %	± 7 %	±1·8 %	±3·0 %

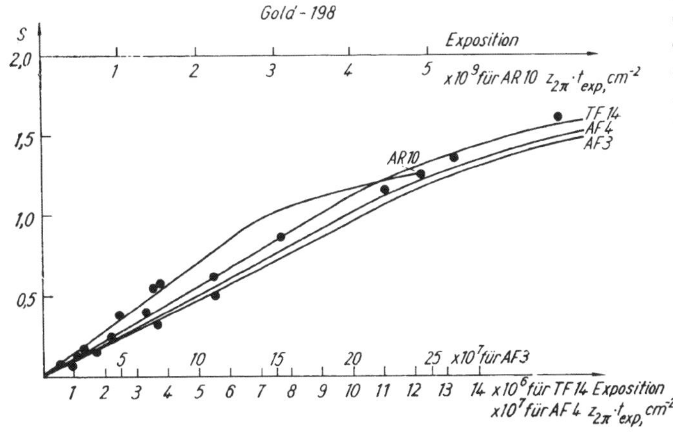

Fig. 4. Density curve (particle fluency per unit area vs. optical density) of autoradiographic emulsions for gold-198

Tab. 3. Results of the determination of the gold concentration in sample No. II by means of gold coated standard No. I

No.	ORWO-AF 3		ORWO-TF 14	
	calc. gold content ng/cm²	deviation from the measured content, %	calc. gold content ng/cm²	deviation from the measured content, %
1	7·63	–21·9	7·73	–20·3
2	8·10	–14·8	8·77	– 6·0
3	8·10	–14·8	9·12	– 2·0
4	6·72	–38·4	8·94	– 4·0
5	6·78	–37·2	9·30	± 0
6	9·43	1·4	10·03	7·3
7	10·33	10·0	–	–
8	8·56	– 8·6	–	–

Tab. 4. Relative deviations of the calculated gold contents from the measured content

film type	maximum deviation of a single measurement %	deviation of the mean value of 8 (AF 3), 6 (TF 14) measurements %
ORWO-AF 3	38·4	11·0
ORWO-TF 14	20·3	3·4

The gold concentration of the two gold coated standards was determined by relative radioactivity measurement:

$$\text{I}\ : c_{Au} = 7{\cdot}63\ \text{ng/cm}^2$$

$$\text{II}: c_{Au} = 9{\cdot}30\ \text{ng/cm}^2$$

The gold coated standard No. II is used as a sample and its gold content is determined autoradiographically by means of the standard No. I. The results are represented in Tab. 3. The highest relative deviation of a single measurement in this investigation is about 38% for ORWO-AF 3 and about 20% for ORWO-TF 14 autoradiographic film. The relative deviations of the mean value of the measurement of gold concentration by means of eight (ORWO-AF 3) and six (ORWO-TF 14) autoradiograms are lower (Tab. 4). The error of an autoradiographic measurement of element concentration is lower than $\pm 10\%$. It depends on the necessary accuracy and on the time available for the investigation.

Conclusion

A method for the standardized handling and quantitative evaluation of autoradiograms using an autoradiography calibration set (ACS) and suitable standard samples of comparison has been represented and demonstrated by an experimental example. By this example it has been shown that it is possible to determine autoradiographically the element concentration in a sample with an error lower than $\pm 10\%$.

Acknowledgements

The authors are grateful to Dr. *J. Flachowsky*, Dr. *H.-P. Rudolph* and Mrs. *A. Lange* for the preparation and chemical etching of the gold-diffused samples, to Mrs. *Ch. Petzold* for carefully making the autoradiograms and to Mrs. *H. Lindemann* for the photometric measurements of optical densities.

Received December 19, 1984
Accepted in revised form February 25, 1985

References

[1] *S. R. Pelc*, Lab. Invest. 8 (1959 127

[2] *W. Herrmann, G. Hartmann, H. Baumbach*, Atompraxis 14 (1968) 8, S. 1

[3] *H. C. Treutler, K. Freyer*, Proc. 13th Int. Symp. Autoradiography, Tábor, CSSR, 3.–5. may 1983, CzAEC, Prague, pp. 57–62

[4] *H. C. Treutler, K. Seifert, G. Schindler, R. Siegel, K. Freyer, K. Wetzel, H. Koch*, Isotopenpraxis 20 (1984) 458

[5] *H. C. Treutler, D. Fischer, K. Freyer*, paper presented at the 14th Int. Symp. Autoradiography, Reinhardsbrunn, GDR, 19–23 November 1984

[6] *H. C. Treutler, K. Freyer, J. Dubnack*, J. Radioanal. Chem. 78 (1983) 155

[7] *H. C. Treutler, K. Freyer*, Isotopenpraxis 19 (1983) 202

Isotopenpraxis 21 (1985) 11, S. 404

Tagungsberichte

G.-v.-Hevesy-Kolloquium am 10. April 1985 in Dresden

G. v. Hevesy ist bekannt als Begründer der Radioanalytik und der Nuklearmedizin und als Entdecker des Hafniums. Er leistete bedeutende Beiträge zur Entwicklung der Röntgenfluoreszenzanalyse, Isotopentrennung, natürlichen Radioaktivität, Geochemie und Biochemie. Zu Ehren seines 100. Geburtstages (1. August 1985) veranstalteten die Arbeitsgemeinschaften Nukleare Analysenmethoden, Radiochemie und Geschichte der Chemie in der Chemischen Gesellschaft der DDR am 10. April 1985 in Dresden ein Kolloquium.

Nach der Eröffnung durch *G. Ackermann* (Freiberg) gab *W. Göbel* (Dresden) einen Überblick über die Biographie von *G. v. Hevesy*. In Beiträgen von *R. Dreyer* (Dresden) über radiometrische Untersuchungen am Astat, *W. Schwarz* (Greifswald) Alkalimetall-Seltenerdmetall-Orthophosphate, eine neue Stoffklasse für Laser, Luminophore und Ionenleiter, *K. Wetzel* (Leipzig), Trennung und Anwendung stabiler Isotope, *H. Rausch* (Budapest), Aktivierungsanalytische Arbeiten im Zentralinstitut für Physik der Ungarischen Akademie der Wissenschaften, *S. Niese* (Rossendorf), die Nachweisgrenze aktivierungsanalytischer Bestimmungen, *R. Münze* (Rossendorf), Indikatorenentwicklung von Hevesy bis zur Doppelkompartmentmarkierungstechnik und *W.-G. Franke* (Dresden), das Konzept von *Hevesy* in der Nuklearmedizin aus heutiger Sicht wurde der gegenwärtige Stand von Arbeitsgebieten vorgestellt, die von *Hevesy* begründet wurden. Dabei wurde die langandauernde Tragfähigkeit seiner Grundideen deutlich.

S. Niese

Eingegangen am 5. 6. 1985

Isotopenpraxis 21 (1985) 11, S. 404 bis 408

Buchbesprechungen

Isotope Hydrology 1983
Proc. of a Symp. on Isotope Hydrology in Water Resources Development, Vienna, 12.–16. September 1983

Vienna: IAEA 1984. 873 S., div. Abb. und Tab. (Proc. Series, STI/PUB/650); karton. ÖS 1680,–

Der Vortragsband des gemeinsam von der IAEA und der UNESCO durchgeführten 6. Symposiums über Isotopenhydrologie umfaßt 41 vollständige Vortragsmanuskripte sowie 30 Posterbeiträge in Form von teilweise ausführlichen Abstracts. Die gehaltenen Vorträge überstreichen ein breites Spektrum der Anwendungen von Isotopentracern in der hydrologischen Praxis. Obwohl die Umweltisotope in den dargestellten Untersuchungen den Vorrang einnehmen, zeigen auch mit künstlichen und insbesondere Radiotracern erhaltene Ergebnisse vielfältige Anwendungsmöglichkeiten im Feldversuch. Zu erwähnen wären beispielsweise Untersuchungen zum Abfluß bzw. zum Studium von Mischungsprozessen in Wasserläufen (z.B. *Dincer* u. a.), zur Lecksuche in Staubauwerken (z.B. *Owczarczyk* u. a.) oder auch zum Transport- und Sedimentationsverhalten von festen Partikeln in offenen Gewässern (z. B. *Gonzalez* u. a.).

Hinter den meisten der vorgestellten Beiträge sind bedeutende ökonomische bzw. wirtschaftliche Zielstellungen erkennbar. Diese reichen von ingenieurhydrologischen Fragestellungen über Probleme der Neubildung und Kommunikation von Wässern bis hin zur großregionalen Bewertung von Wasserressourcen. Auf Grund der Kompliziertheit der Untersuchungsobjekte und des Informationsgewinnes durch korrelative Betrachtungen sind heute komplexe Isotopenanwendungen (meist mindestens ^{2}H, ^{3}H, ^{18}O, ^{13}C, ^{14}C) die Regel. Bei ^{39}Ar wird oftmals schlechte Übereinstimmung mit ^{14}C-Altern gefunden (*Geyh* u. a., *Forster* u. a.). Die Bedeutung von ^{85}Kr für die sichere Jungwassereinstufung ist unumstritten, die international verfügbare Analysenkapazität immer noch gering. Erreicht wurde (*Michelot* u. a.) die Messung ^{36}Cl ($T_{1/2} = 301\,000$ a) als geochemischer Tracer oder Datierungsnuklid. Der tatsächliche Aussagewert für den nichtglazialen Bereich muß abgewartet werden.

Die Isotopenhydrologie ist nicht zuletzt durch die erreichten Meßgenauigkeiten in der Lage, bedeutende Informationen über Genese, Zusammensetzung, Kommunikation, Fließverhalten, Transitzeiten oder die Bilanz von Wasserreservoirs zu geben. Gleichzeitig unterliegt sie selbst einer intensiven Bearbeitung in ihren methodischen Voraussetzungen und Mitteln. So wurden z.B. von *Evans* u. a. ^{14}C-Konzentrationen in sekundären Kalziten eines Sandsteinaquifers von bis zu 20% modern nachgewiesen. Die Isotopenzusammensetzung des Gesteinskarbonates ist eine wichtige Größe in Modellrechnungen zur Ermittlung des

^{14}C-Anfangswertes. Neben derartigen hydrochemischen Prozessen im Aquifer wurden in einigen Arbeiten die Vorgänge in der ungesättigten Bodenzone unter verschiedenen geomorphologischen und klimatischen Bedingungen (Aspekt Grundwasserneubildung) betrachtet. In einer umfangreichen Arbeit über Isotopenvariationen in Luftfeuchte und Niederschlag wurden u. a. bei Spektralanalysen der Tageswertabfolge für die Stationen Heidelberg und Krakau Periodizitäten von 50–67 d, 29–36 d, 20–22 d und 10 bis 11 d nachgewiesen.

Die mit dem Tagungsband vorliegende Sammlung vorwiegend praktischer Ergebnisse und Erfahrungen bietet einen guten Überblick über die gegenwärtig typischen Anwendungsziele, Methoden und Probleme der Isotopenhydrologie. *R. Trettin*

Klassifikation von Lagerstättenvorräten mit Hilfe der Geostatistik

Wiss. Red.: W. Ehrisman u. H. W. Walther: Weinheim — Deerfield Beach (Florida) — Basel: Verlag Chemie 1983. 164 S., div. Abb. u. Tab. (Schriftenreihe der Ges. Dt. Metallhütten- und Bergleute, Heft 39); karton. DM 60,—

Schon seit einigen Jahrzehnten ist man sich auch in den westlichen Ländern darüber klar, daß die Einteilung der Vorräte von mineralischen Rohstoffen in „sichere", „wahrscheinliche", „mögliche" usw. recht unscharf ist und deshalb einer Präzisierung bedarf. Besonders deutlich wird dieser Mangel empfunden, wenn die Vorräte solcher Metalle wie Uranium ausgewiesen werden sollen, da sich die Lagerstätten oftmals durch starke räumliche Schwankungen der Gehalte und Mächtigkeiten auszeichnen — gar nicht zu sprechen von Edelmetallen wie Silber und Gold, deren Erkundung und Vorratsberechnung selbst erfahrene Fachleute immer wieder vor schwierige Entscheidungen stellt. In den sozialistischen Ländern hat man — nach dem Vorbild der Sowjetunion — Klassifikationen eingeführt, in denen die (quantifizierbare) Dichte der verfügbaren Aufschlüsse eine wichtige Rolle spielt; daneben wird der (nicht quantifizierbare) Grad der wissenschaftlichen Auswertung der geschaffenen Aufschlüsse zur Einstufung der Vorräte in die einzelnen Klassen herangezogen. Dieses Vorgehen vermeidet die Verwendung solcher Adjektive, wie sie eingangs genannt wurden. Es ist aber klar, daß auch hier den einzelnen Klassen eine bestimmte Unsicherheit oder ein Fehler eigen ist, den man auf verschiedene Weise abzuschätzen versucht.

Einen ganz anderen Weg zur Lösung dieser Aufgaben bietet eine Methode, die unter dem Namen Geostatistik bekannt wurde. Sie geht auf Arbeiten von *Sichel*, *DeWijs*, *Krige* und anderen Autoren zurück, die sich etwa seit 1945 mit der Berechnung von Vorräten in sehr absätzigen Lagerstätten beschäftigten (vornehmlich mit den südafrikanischen Golderzen). Später griffen *Formery*, *Ziegler* und vor allem *Matheron* diese Arbeiten auf und entwickelten eine Theorie, die im wesentlichen auf der bekannten Theorie der (stationären) Zufallsfunktionen beruht, aber einen mathematisch vereinfachten Apparat „für den Hausgebrauch" anbietet. Mit Hilfe der Geostatistik kann die Mehrzahl der Risikofaktoren numerisch abgeschätzt werden — ein bedeutender Fortschritt für die Vorratsberechnung („Der Techniker interessiert sich für das Ergebnis einer Messung, der Physiker für ihren Fehler").

Um einen größeren Kreis von Fachleuten mit den Grundlagen der Geostatistik bekannt zu machen, hat die Gesellschaft Deutscher Metallhütten- und Bergleute (BRD) im Jahre 1982 eine Diskussionstagung veranstaltet, deren Vorträge in der vorliegenden Broschüre veröffentlicht sind. Die einzelnen Beiträge sind gut geeignet, den gewünschten Ein- und Überblick zu vermitteln. Besonderes Interesse verdienen einige Anwendungsbeispiele, aus denen das methodische Vorgehen abgelesen werden kann (Uraniumerze, Steinkohlen, Rohstoffe (Erzschlämme) am Meeresboden). Im Anhang wird eine Klassifikation von 1959 abgedruckt, wodurch der inzwischen erzielte Fortschritt erkennbar wird. Weiterhin ist eine Wortliste (deutsch — englisch — französisch) beigefügt. Ein Sachregister schließt das Werk ab. Man vermißt die Wiedergabe der Diskussionen zu den Vorträgen.

Die Veröffentlichung ist zweifellos geeignet, die Methoden der Geostatistik einem großen Kreis von möglichen Anwendern bekannt zu machen und ihre Erprobung unter verschiedensten Bedingungen zu fördern. *K. H. Köhler*

Die Physik der Zeitrichtung

H. D. Zeh. Berlin/Heidelberg/New York/Tokyo: Springer-Verlag 1984. V, 86 S., mehrere Abb. (Lectures Notes in Physics, Vol. 200) kart. DM 18,—; US \$ 7.10

Der vorliegende Band der Lecture Notes ist dem nach wie vor hochaktuellem Problem der Existenz einer ausgezeichneten Zeitrichtung (des sog. Zeitpfeiles) in den Naturphänomenen gewidmet. Untersucht werden Klassen von Vorgängen, die bei vorgegebenen zeitsymmetrischen mikroskopischen Grundgleichungen eine Zeitrichtung auszeichnen, wie z.B. Wellenphänomene (Elektrodynamik), die Thermodynamik in ihrem Zusammenhang mit der statistischen Mechanik, der quantenmechanische Meßprozeß, gravitierende Systeme u. a.

Nach einer ausführlichen Diskussion der Frage, warum in der Natur vornehmlich die retardierten (und nicht die avancierten) Lösungen der *Maxwell*schen Gleichungen beobachtet werden, wendet sich der Autor dem statistisch mechanischen Zugang zur Erklärung des thermodynamischen Zeitpfeiles zu. Dabei wird, zunächst im Rahmen der klassischen Statistik, das Konzept der Mastergleichungen (Zwanzig) diskutiert und der durch Projektion entstehende Informationsverlust in Zusammenhang mit dem thermodynamischen Zeitpfeil gebracht. Interessante Ausführungen finden sich auch zum Problem der Objektivierung des Relevanzkonzeptes und zum Zusammenhang zwischen Thermodynamik und Informationstheorie.

Als besonders wichtig und interessant erscheint die Untersuchung der möglichen Verknüpfung zwischen den Problemen der irreversiblen Thermodynamik und Statistik und dem Interpretations- bzw. Meßproblem der Quantenmechanik. Es wird dabei die Frage des Entropiezuwachses bei der Projektion eines Zustandes bzw. bei dessen Messung infolge des Kollapses der Wellenfunktion untersucht.

Die Weiterführung dieser Gedanken führt dann notwendigerweise auf kosmologische Fragen, wobei vornehmlich die „unwahrscheinliche Anfangsbedingung" (Urknall) und die Thermodynamik schwarzer Löcher diskutiert werden. Die endgültige Antwort auf die Frage nach dem Urzeitpfeil erfordert aber, wie abschließend klar herausgestellt wird, eine Lösung des Problems der Quantengravitation und der Quantisierung der Zeit.

Das Buch ist trotz der Kompliziertheit des dargestellten Stoffes außerordentlich anschaulich geschrieben und kann als Einführung für Interessenten sehr empfohlen werden, es dürfte aber auch für Spezialisten eine Reihe von interessanten und neuartigen Gedanken bieten. *R. Der*

Operational Safety of Nuclear Power Plants

Proceedings of an International Symposium. Vienna: IAEA 1984. Vol. I: 380 S., div. Abb. u. Tab.; Vol. II; 456 S., div. Abb. u. Tab.; kart. ÖS 750.– u. 900.–

Mit den beiden Bänden "Operational Safety of Nuclear Power Plants" liegt die Berichterstattung über ein IAEA-Symposium vor, das im Mai 1983 in Marseilles zur Kernkraftwerkssicherheit stattgefunden hat. Die Veranstaltung vermittelte einen Überblick über den derzeit erreichten Wissensstand, dem Erfahrungen aus etwa 3000 Reaktorbetriebsjahren zugrunde liegen. Da sich in Zukunft sowohl die Anzahl der KKW-Anlagen als auch die Zahl der Länder, in denen KKW betrieben werden, weiter erhöhen wird, war es Ziel des Symposiums, die gesammelten Erfahrungen zu verbreiten.

Wie komplex das Gebiet der KKW-Sicherheit ist, wird durch die Themen der sieben Sektionen deutlich, zu denen vorgetragen und diskutiert worden ist. In der Sektion I „Maßnahmen für den sicheren Betrieb" wurden die von der IAEA entwickelten Richtlinien zur Störfallberichterstattung diskutiert, durch die Verbesserungen in der internationalen Zusammenarbeit angestrebt werden. Es wird vorgeschlagen ein Berichtssystem über Störfälle aufzubauen, Expertengruppen mit der sachkundigen Berichterstattung über das KKW Betriebsgeschehen zu betrauen und Trainingsprogramme zu entwickeln, die auch Empfehlungen für Maßnahmen in Störfall-Situationen enthalten. Weiterhin wurden Verfahren der kontinuierlichen Prozeß- und Anlagenüberwachung, Methoden der Schadensfrüherkennung und Verfahren für den Qualitätsnachweis vorgestellt, die sowohl der Sicherheit dienen als auch die Anlagenverfügbarkeit günstig beeinflussen. Anliegen der Sektion II „Leitung und Schulung des Personals" war es, die Forderung nach qualifiziertem Anlagenpersonal zu unterstreichen. Es wird als notwendig erachtet, das Betriebspersonal ständig zur Unterstützung von Sicherheitsmaßnahmen zu motivieren. Zu Recht wurde darauf verwiesen, daß sich ein qualifiziertes Personal auch bei der Gewährleistung anspruchsvoller Instandhaltungsaufgaben bewährt.

Die Sektion III beschäftigte sich mit Problemen der „Mensch-Maschine-Kommunikation". Unter anderem wurden Möglichkeiten zur Reduzierung menschlichen Fehlverhaltens diskutiert. Mit dem Ziel, Betriebsvorschriften zu entwickeln, die den Operator befähigen, auch gegenüber unvorhergesehenen Ereignissen gerüstet zu sein, wurden typische Verhaltensweisen des Betriebspersonals zur Anlage oder zu verwaltungstechnischen Verfahren analysiert. Auch Beiträge, die Anregungen zu Kontrollraumgestaltung enthielten, beschäftigten sich mit ergonomischen Aspekten. Ein weiterer Themenkomplex dieser Sektion galt Diagnosesystemen zur Ermittlung des Anlagenzustandes. Dabei konnten Fortschritte hinsichtlich der aus der Diagnose abgeleiteten Handlungsvorschriften für den Operator festgestellt werden. Ferner wurde erörtert, ob es zweckmäßig ist, den Verantwortlichen verschiedener Leitungsebenen zu ermöglichen, voneinander unabhängige Anlagendiagnosen vornehmen zu können.

In der Sektion IV „Technische Verbesserungen für die Betriebssicherheit" wurde berichtet, daß Forschungsarbeiten gegenwärtig vor allem denjenigen Komponenten gelten, die als Schutz gegen Überdruck wirken, sowie Prozeß- und Anlagenbedingungen, die zu einer Störung geführt haben. Entwickelt wurden auch Verfahren zur Eigendiagnose von Detektoren, durch die der Operator über den Wahrheitsgehalt der angezeigten Daten informiert wird.

Die Sektion V „Auswertung von Betriebserfahrungen" beschäftigte sich hauptsächlich mit Erfahrungen, die in verschiedenen Kernkraftwerken mit den Sicherheitskomponenten gesammelt worden sind. Es erweist sich jedoch als schwieriger als ursprünglich angenommen, diese Erfahrungen zu verallgemeinern. Besonderes Interesse galt den Prozessen, die zur Bildung einer Dampfblase im Druckgefäß führen und die ausreichende Kühlung der Brennelemente gefährden können. Auch hier wurde auf die Wechselwirkungen hingewiesen, die zwischen einer sicheren und ökonomisch effektiv arbeitenden Anlage bestehen.

Die Sektion VI „Internationale Bemühungen um eine Verbesserung der Sicherheit" erörterte Ergebnisse, die beim Aufbau eines internationalen Störfallberichtssystems bisher erreicht worden sind und schlug Richtlinien für die Qualifizierung von Operatoren vor.

In den Vorträgen der Sektion VII „Vorbereitung auf außergewöhnliche Ereignisse" wurde die Notwendigkeit unterstrichen, entsprechende Vorkehrungen zu treffen. Diskussionsgegenstand war hauptsächlich der separate Kontrollraum, der im Fall außergewöhnlicher Ereignisse hinreichend Sicherheit bietet, eine Fernüberwachung von Störfällen, die Entscheidungen übergeordneter Institutionen ermöglicht und ein speziell ausgelegtes System, das dem Operator Informationen über den Anlagenzustand vermittelt.

Außerdem bot das Symposium Gelegenheit, zu den Komplexen „Mensch-Maschine-Kommunikation" sowie „Zukünftige Maßnahmen zur Verbesserung der Betriebssicherheit" im Plenum zu diskutieren. Bei dieser Gelegenheit wurden auch die Leistungen gewürdigt, durch die die Kernkraftwerkssicherheit in den vorangegangenen Jahren kontinuierlich erhöht worden ist. *H.-F. Brinckmann*

Irradiation Technology
Proceedings of an International Topical Meeting, Grenoble, France, September 28–30, 1982

Hrsg.: P. v. d. Hardt u. H. Röttger. Dordrecht/Boston/Lancaster: D. Reidel Publishing Company 1983, 759 S. (engl./franz./deutsch); geb. Dfl. 230.–; US $ 100.–

Der Titel „Bestrahlungstechnologie", der dieser von der European Nuclear Society und anderen durchgeführten Veranstaltung gegeben wurde, will besagen, daß die Forschungsgeräte, die bei der Verwendung von Neutronen aus Forschungsreaktoren für Materialtestzwecke zum Einsatz gelangen, in ihrem technischen und methodischen Aufbau sowie in ihrer jeweiligen Zweckbestimmung dargestellt werden. Diese Geräte sind in ihrer Ausführung modernen kernenergetischen Ansprüchen angepaßt, oftmals sehr teuer und aufwendig in Entwicklung und Betrieb. Sie wurden vor dieser Veranstaltung letztmalig 1973 (Jackson, Wyoming) vorgestellt und diskutiert und bildeten 1982 den Gegenstand von gleich 2 internationalen Konferenzen

– der ANS-Konferenz (Salt Lake City, Utah, April 1982) und
– der hier charakterisierten Konferenz von Grenoble,

wobei letztere eine stärkere Betonung auf thermische Kernreaktoren legt. Die Beiträge stammen bevorzugt aus kernenergetischen Forschungseinrichtungen Großbritanniens, Frankreichs, der Bundesrepublik Deutschland und der Niederlande, mehr sporadisch aus den USA und Japan.

Als Ordnungsprinzip für die Beiträge zu dieser Konferenz bietet sich neben der Zuordnung zu Leichtwasser-, schnellen und gasgekühlten Reaktoren eine Charakterisierung der Zweckbestimmung der Forschungsgeräte für Bestrahlungstechnik an. So dienen solche Geräte (Schleifen, Sonden, Kapseln in Verbindung mit ihrer Instrumentierung) an Leichtwasserreaktoren vorwiegend der Untersuchung des Verhaltens von Brennstoffen unter stationären/nichtstationären Betriebsbedingungen der Reaktoren oder bei Störfällen. Weiter werden Geräte erläutert, die der Untersuchung von Strukturmaterialien, besonders Druckbehälterstählen, zugedacht sind und sowohl während als auch nach der Bestrahlung zum Einsatz gelangen.

Analoge Zweckbestimmungen bestehen bei der Bestrahlungstechnik für schnelle und gasgekühlte Reaktoren. Auch hier dominieren Untersuchungen des Strahleneinflusses auf Brennstoffe (Uran-Plutonium-Mischoxide, vibriert oder in Kugelgestalt), auf Hüllenmaterialien (austenitischer Stahl) und auf instrumentierte Brennelementkassetten unter definierten Betriebsbedingungen bis hin zu Störfalluntersuchungen.

Aus einem Abschnitt, mit dem Begriff Verschiedenes überschrieben, sind ein Rückblick auf die Nutzung des Forschungsreaktors FR 2 (BRD), Brennelemente für Forschungsreaktoren, Dotierungsuntersuchung von Phosphor und das moderne Verfahren der Brennelementtomographie hervorzuheben. *H. Steinkopff*

International Tokamak Reactor, Phase Two A, Part I Report of the International Tokamak Reactor Workshop organized by the IAEA held in seven sessions in Vienna during 1981—1983

Vienna: IAEA 1983. 772 S., div. Abb. u. Tab. (Panel Proc. Series, STI/PUB/638); kart. ÖS 1520.—

Es liegt jetzt der dritte Band der Serie über den „International Tokamak Reactor Workshop" vor. Der internationale Workshop war 1978 von der IAEA ins Leben gerufen worden mit dem Ziel, die Entwicklung eines thermonuklearen Fusionsreaktors zu fördern, die neuesten Erfahrungen der international führenden Kernfusionsforschungszentren zu nutzen und mit kooperativen Bemühungen ein Projekt für den nächsten Schritt auf dem Wege zu einem Tokamak-Reaktor zu schaffen. Im Juli 1981 wurde die Phase 2A begonnen mit der Aufgabe, kritische technische Probleme, die sich während der Bearbeitung der Phase 1 herausgestellt hatten, im Detail zu untersuchen und konkrete Lösungswege vorzuschlagen. An der Lösung der Aufgaben der Phase 2A sind zahlreiche Expertengruppen in den beteiligten nationalen Einrichtungen bis 1983 tätig gewesen. Ihre Ergebnisse wurden in sechs Workshop-Veranstaltungen abgestimmt, und die Ergebnisse dieser Arbeitsperiode des INTOR-Workshops sind im vorliegenden Band zusammengefaßt.

Der Phase-2A-Bericht ist folgendermaßen gegliedert: Nach einer kurzen Einleitung (Kap. I) wird im Kapitel II ein Überblick über die in der Phase 2A durchgeführten Arbeiten und eine Zusammenfassung der Ergebnisse und Empfehlungen gegeben. Kapitel III beschreibt den Stand des INTOR-Konzepts nach Berücksichtigung sämtlicher Veränderungen und Korrekturen, die aus der Phase 2A resultierten. Die Kapitel IV—XII enthalten die Ergebnisse der Untersuchungen der Phase 2A im Detail: Kap. IV — Plasmahalterung und -steuerung; Kap. V — HF-Zusatzheizung und nichtinduktive Plasmastromerregung; Kap. VI — Physik der Verunreinigungssteuerung; Kap. VII — Verunreinigungssteuerung und Konstruktion der ersten Wand; Kap. VIII — Tritium und Blanket; Kap. IX — Magnetsystem; Kap. X — Elektromagnetische Wechselwirkung des Plasmas mit dem Tokamaksystem; Kap. XI — Mechanische Konfiguration; Kap. XII — Technische Prüfungen und Voruntersuchungen.

Die beiden nächsten Kapitel sind Kosten-Risiko-Nutzen-Untersuchungen und dem Zeitplan gewidmet. Im Kapitel XV werden die F/E-Forderungen resümiert, gegliedert nach Plasmaphysik, Kernphysik, Technik und Technologie. Im Kapitel XVI werden die Konstruktionsmerkmale der INTOR-Anlage zusammengefaßt.

Das Buch ist in erster Linie wertvolle Informationsquelle und Nachschlagewerk für Fachleute, die sich mit dem Problem der Kernfusionsforschung mit magnetischer Plasmahalterung befassen. Es ist aber auch von großem Wert für alle, die sich eine Vorstellung vom gegenwärtigen Stand der Tokamakentwicklung verschaffen möchten, denn der Kenntnisstand und die offenen, noch zu lösenden Probleme werden deutlich herausgestellt. *H. Wolff*

Proceedings of the IAEA Consultants Meeting on Nuclear Data for Bore-Hole and Bulk-Media Assay Using Nuclear Techniques; Krakow, Poland, 14—18 November 1983

Ed.: K. Okamoto. Vienna: IAEA 1984. 348 S., div. Abb. u. Tab. (INDC(NDS)-151/L)

Zur schnellen und empfindlichen Bestimmung von chemischen Elementen in Gesteins-, Mineral- und Erzproben werden kernphysikalische Methoden in wachsendem Maße eingesetzt. Ihre Vorteile sind hinreichend groß, um die klassischen Verfahren der Analytik ganz oder wenigstens teilweise bei bestimmten Anwendungsfällen zu verdrängen. Neben der geologischen Erkundung ist die Prozeßkontrolle in der Aufbereitung und Verarbeitung der gewonnenen Rohstoffe ein wichtiges Anwendungsgebtiet geworden, in dem die kernphysikalische Analyse einen beträchtlichen qualitativen und quantitativen (in Geld ausdrückbaren) Nutzen bringt. Im gleichen Maße, wie die Anwendungsbreite zunimmt, werden mehr und mehr kompliziertere Aufgaben gestellt: Unterscheidung von Elementen mit ähnlichen kernphysikalischen Eigenschaften, Verkürzung der Meßzeiten, Verbesserung der Nachweisgrenzen bis in den technogenen Umweltbereich usw. Diese wachsenden Anforderungen an die Analytik bedingen eine bessere Kenntnis der kernphysikalischen Daten, weshalb das INDC (International Nuclear Data Committee) ein besonderes Treffen über diesen Gegenstand veranstaltet hat. An dem Treffen nahmen 27 Spezialisten aus 12 Staaten teil. Es wurden 8 Übersichtsvorträge und 15 Vorträge in 5 Sektionen gehalten. Die hauptsächlichen Themen und Probleme des Treffens waren:

- Übersichten des gegenwärtigen Standes und der Anforderungen an die Kerndaten, die für die Analytik bei der Erkundung, Gewinnung und Verarbeitung von Öl, Gas, Kohle, Erzen und anderen Rohstoffen benötigt werden;

— Erörterung der Ursachen von Unsicherheiten und ihrer Auswirkungen bei solchen Methoden wie Bohrlochmessungen, Gammaspektrometrie, Röntgen-Fluoreszenz-Analysen, Aktivierungsanalysen usw.,
— Auswahl von Methoden, die für Entwicklungsländer besonders zu empfehlen sind;
— Formulierung von Empfehlungen für die künftige Präzisierung und Vertiefung unserer Kenntnis der Kerndaten.

Die vorliegenden Proceedings bestätigen, daß dieses Treffen seinem Anliegen gerecht wurde und enthalten zahlreiche Informationen. Als Beispiele mögen herausgegriffen werden: Kerndaten für Bohrlochmessungen auf Öl und Gas (*Allen* u. a., USA); Anwendungen im Untertage-Bergbau (*Przewlocki*, Polen); Grundlagen der radiometrischen Bohrlochmessungen (*Kozhevnikov*, UdSSR); Nachweis von Edelmetallen durch Bohrlochmessungen (*Senftle* u. a., USA); Bestimmung von Silizium und Aluminium (*Divos* u. a., Ungarn); Untersuchungen mit *Monte-Carlo*-Methoden zur Gesteinsdichtebestimmung (*Clayton* u. a., UK); Flugzeug-Aufnahmen (*Wendt* u. a., BRD); Bohrloch-Modelle für K–U–Th (*Arnold*, USA).

Spezialisten der geophysikalischen und geologischen Erkundung und der Aufbereitung ebenso wie Physiker, die auf diesen Gebieten tätig sind, werden die Proceedings mit Interesse für ihre Arbeit auswerten.

K. H. Köhler

Nuclear Winter
The Human and Environmental Consequences of Nuclear War

M. A. Harwell. Berlin — Heidelberg — New York — Tokyo: Springer-Verlag 1984. XXI + 179 S., 27 Abb.; Hard cover DM 54,—

Das Buch gibt die erste umfassende Darstellung der Konsequenzen eines Kernwaffenkrieges. Es behandelt sowohl die Auswirkungen der Druck- und Hitzewellen, der Sofortstrahlung und des lokalen und globalen Fallout radioaktiver Stoffe, die von Kernwaffenexplosionen ausgelöst werden, als auch die langfristigen Veränderungen des globalen Klimas, die erst in den letzten 2—3 Jahren Gegenstand umfangreicher Untersuchungen sowjetischer wie amerikanischer Wissenschaftler geworden sind. Dabei wird von einer Vorstellung über den Ablauf eines Kernwaffenkrieges ausgegangen, die sich an jenes Scenario anschließt, das von der Ambio Advisory Group (Reference Scenario: how a nuclear war might be fought. Ambio **11** (1982) 94—99) ausgearbeitet wurde: Rasche Eskalation zu einem globalen Kernwaffenkrieg, in dem beide Seiten etwa die Hälfte ihres heutigen Kernwaffenpotentials zur Wirkung bringen. Wie die von der Zeitschrift Ambio initiierten Untersuchungen gezeigt haben, würde ein solcher Krieg kurzfristige Wirkungen auslösen, die den Tod eines erheblichen Teils der Menschheit, jedenfalls von mehr als einer Milliarde Menschen, zur Folge hätten. Waldflächen etwa von der Größe Skandinaviens würden niederbrennen. Die dabei entstehenden Rußteilchen und die mit den Druckwellen der Kernwaffenexplosionen emporgeschleuderten Staubteilchen würden lange Zeit in der Atmosphäre bzw. Stratosphäre verweilen und den Wärmehaushalt der Erde in zweifacher Weise verändern: Reflexion eines Teils der auf die Erde eindringenden Sonnenstrahlung und, da diese Reflexion überwiegend in den höheren, also kein CO_2 enthaltenden Schichten der Atmosphäre erfolgen würde, Minderung des Treibhauseffektes.

Die genauere Untersuchung dieser Effekte hat gezeigt, daß zumindest die Nordhalbkugel sich mit hoher Wahrscheinlichkeit um etwa 30 °C abkühlen und daß dieser „nukleare Winter" womöglich auch auf die südliche Hemisphäre übergreifen würde. Dies aber hätte nicht nur den Untergang der menschlichen Zivilisation zur Folge, sondern könnte auch zum Untergang der gesamten Menschheit und zum Aussterben vieler Tier- und Pflanzenarten führen.

Der Autor leitet die Untersuchungen über die Auswirkungen auf die Landwirtschaft im Rahmen des Projektes „Auswirkungen von Kernwaffenkriegen auf die Umwelt" des International Council of Scientific Unions. Er beschreibt den heutigen Stand unserer Kenntnisse über die Wirkungen mit Kernwaffen geführter Kriege sowohl exakt als auch in einer für jeden naturwissenschaftlich gebildeten Menschen verständlichen Form. Die Darstellung ist erstaunlich leidenschaftslos. Der Autor enthält sich jeglicher Aussagen über die politischen Konsequenzen aus der grausigen Vision, die das Buch entwirft. Trotzdem wird es jeden aufrütteln, der es liest. "None of us can responsibly sit back today and let somebody else worry about this problem", heißt es im Vorwort von *R. W. Peterson* zu dem lesenswerten Buch.

K. Wetzel

Homogene Katalyse durch Übergangsmetalle

C. Masters. Übers. aus dem Engl. Ju. I. Ermakova, V. A. Semikolenova. Moskau: Mir 1983. 300 S., zahlr. Abb. und Tab. (in Russisch), kart. M 13.50

Die russische Übersetzung des 1981 als erste Auflage in englischer Sprache erschienenen Buches liefert einen interessanten Überblick über die rasche Entwicklung der metallorganischen Komplexkatalyse in den letzten 30 Jahren. In klarer und übersichtlicher Form gibt der Autor eine konzentrierte Einführung in dieses sich rasch entwickelnde Gebiet der Chemie. In 4 Kapiteln werden neben einigen allgemeinen Grundlagen die folgenden komplexkatalysierten Reaktionen behandelt:

Hydrierung von Alkenen und Alkinen, Alken- und Skelettisomerisierung, Carbonylierung von Alkenen, Alkinen und Alkoholen, Hydroformylierung, Oligomerisation von Alkenen und Dienen, Alkenpolymerisation, Autoxidation von Kohlenwasserstoffen, Epoxidation und Metathese von Alkenen, reduktive Stickstoffixierung und die Alkanaktivierung. Den Abschluß bildet ein kurzer Ausblick auf einige Entwicklungstrends.

Für jede Reaktion werden der katalytische Zyklus, die praktische Bedeutung und gegebenenfalls die industrielle Durchführung sowie die Chancen einer asymmetrischen Katalyse erläutert. Insgesamt werden 448 Originalarbeiten zitiert, aus denen der große Anteil isotopentechnischer Methoden bei der Klärung mechanistischer Details komplexkatalysierter Reaktionen hervorgeht. Sehr nützlich sind zahlreiche Angaben von weiterführender Literatur am Ende jeden Abschnitts, die von den Übersetzern noch durch eine Reihe zusätzlicher Literaturangaben vorwiegend sowjetischer Autoren ergänzt wurden.

Das Buch kann Spezialisten, aber auch allen denen, die sich mit homogener Katalyse befassen wollen als Einführung, Nachschlagewerk und Anregung für weitere Arbeit auf diesem Gebiet empfohlen werden.

S. Rummel

Silicat-Lexikon

Band 1—2

Nichtmetallisch-anorganische Werkstoffe,
vorwiegend silicatische Minerale und Gesteine, Hilfswissenschaften, Biographien
Herausgegeben von Wilhelm Hinz

1985. Ca. 1120 Seiten — 17 × 24 cm — Leinen DDR ca. 150,— M; Ausland ca. 150,— DM
Bestell-Nr. 752 512 0
Bestellwort: Hinz, Silicat-Lexikon 6458

Eine Enzyklopädie der Silicate im weitesten Sinne! Sie enthält sowohl die Werkstoffe Glas, Keramik, Glasuren, Bindebaustoffe als auch ihre Grundlagen, ferner silicatische Mineralien und Gesteine, Untersuchungsmethoden und andere „Hilfswissenschaften" der Silicate sowie Biographien von Silicatwissenschaftlern aus aller Welt. Das Buch ist ein unentbehrliches Nachschlagewerk. Als Anhang des zweibändigen Werkes sind die enthaltenen Begriffe in deutsch, englisch, russisch und französisch aufgeführt.

Interessentenkreis: Naturwissenschaftler, Ingenieure, Techniker

ISOTOPENPRAXIS

Band 21 | 1985 | Heft 11

Contents | Inhalt

►Language of Publication | ►Publikationssprache

Isotopenpraxis is indexed in Current Contents/PC & ES, Science Citation Index, and INIS Atomindex

FSC
www.fsc.org
MIX
Papier | Fördert gute Waldnutzung
FSC® C083411

Kolibri 360 GmbH
Ferdinand-Jühlke-Straße 7
99095 Erfurt, Deutschland
produktsicherheit@kolibri360.de